Human-Intelligence-Based Manufacturing

HUMAN-INTELLIGENCE-BASED MANUFACTURING

Edited by

Y. Ito

With 86 Figures

Springer-Verlag
London   Berlin   Heidelberg   New York
Paris   Tokyo   Hong Kong
Barcelona   Budapest

Yoshimi Ito, Dr.-Eng. C Eng.
Department of Mechanical Engineering for Production
Faculty of Engineering
Tokyo Institute of Technology
2-12-1 Ohokayama, Meguro-ku
Tokyo 152, JAPAN

Series Editor

D. T. Pham
University of Wales College of Cardiff
School of Electrical, Electronic and Systems Engineering
P.O. Box 917, Cardiff CF2 1XH

ISBN-13:978-1-4471-2016-2   e-ISBN-13:978-1-4471-2014-8
DOI: 10.1007/978-1-4471-2014-8

British Library Cataloguing in Publication Data
Human-intelligence-based Manufacturing.—
(Advanced Manufacturing Series)
 I. Ito, Yoshimi II. Series
 670.285
ISBN-13:978-1-4471-2016-2

Library of Congress Cataloging-in-Publication Data
Human-intelligence-based manufacturing / edited by Y. Ito.
    p.   cm.— (Advanced manufacturing series)
  Includes bibliographical references and index.
  ISBN-13:978-1-4471-2016-2 : $106.50 (est.)
  1. Flexible manufacturing systems. 2. Computer integrated manufacturing
systems. I. Ito, Y. (Yoshimi), 1940– . II. Series.
TS155.6.H835   1993
670.42'7--dc20   93-3305

Typeset by Photo·graphics, Honiton, Devon

69/3830-543210 Printed on acid-free paper

Preface

The *Flexible Computer-Integrated Manufacturing Structure* (FCIMS) is a notable innovation in the field of production science and technology. It provides effective and powerful tools which rapidly manufacture both consumer and defence products as required by society. FCIMS is now mature, having completed its initial development, and is embarking on further development to meet the increased diversity of user demand.

New manufacturing systems have been tested in recent trials. In particular, these seek to combine the deep knowledge and experience of engineers, and human factors with the manufacturing technology. Additionally, this new technology may also become applicable to markets for highly personalized goods, such as aesthetic products. The techniques are part of a new global manufacturing culture which may assist in the dissemination of understanding of these new manufacturing technologies. This book both reviews and predicts future trends in these areas, and focuses on industrial sociology in highly automated manufacturing environments, FCIMS of global network type and culture of manufacturing, and also on thought model-based manufacturing. Chapter 1 introduces the material that follows. Chapters 2, 3 and 4 consider the organizational and personnel aspects of highly automated manufacturing environments. Chapters 5 and 6 discuss FCIMS of the global network type. Chapter 7 describes anthropocentric intelligence-based manufacturing and Chapters 8 and 9 the "culture of manufacturing".

There are other well-known studies which have been undertaken in the fields of "industrial culture" and "industrial sociology". These do not generally approach the field from either a scientific or technical viewpoint. This book seeks to help establish a new interdisciplinary field of "socio-mechanical engineering". Its contributors are chosen from Japan, Europe and the USA where leading-edge research in FCIMS and its related technologies is being carried out.

The editor would wish to recommend the book not only to production engineers, enterprise managers and related groups in both academia and industry, but also to sociologists, and technology transfer specialists.

First of all, I would like to offer my sincere thanks to Dr D.T. Pham, Professor of Computer-Controlled Manufacture of the University of Wales College of Cardiff, for kindly giving me the opportunity to edit this book. My thanks are further extended to the contributors for imparting their valuable leading-edge knowledge on the production structure of the next generation, and also to Mr Nicholas Pinfield, Mrs Lynda Mangiavacchi and Mrs Imke Mowbray (all of Springer-Verlag London Limited) for their expert help and painstaking efforts in the production of this book.

Tokyo, 1993   *Yoshimi Ito*

Contents

Contributors

H. Hirsch-Kreinsen

Institut für Sozialwissenschaftliche Forschung e.V.,
Jakob-Klar-Strasse 9,
8000 München 40, Germany

T. Ihara

Department of Precision and Mechanical Engineering,
Chuou University, Kasuga 1-13-27,
Bunkyou-ku,
Tokyo, Japan

J. Iimura

Computer Systems R&D Laboratory,
Omron Co., Tadao 2-5-48,
Machida City,
Tokyo, Japan

Y. Ito

Department of Mechanical Engineering for Production,
Tokyo Institute of Technology,
2-12-1 Ohokayama, Meguro-ku,
Tokyo, Japan

T. Itoh

Overseas Operations Promotion Office,
Hitachi Ltd., 4-6 Kanda-surugadai,
Chiyoda-ku,
Tokyo, Japan

M.J. Kolar

Mechanical Engineering Department,
School of Engineering, University of Pittsburgh,
Pittsburgh, PA 15261, USA

C. Köhler

Institut für Sozialwissenschaftliche Forschung e.V.,
Jakob-Klar-Strasse 9,
8000 München 40, Germany

L. Mårtensson

Department of Work Science,
Royal Institute of Technology,
S-100 44 Stockholm, Sweden

N. Mårtensson

Department of Production Engineering,
Chalmers University of Technology,
S-412 96 Göteborg, Sweden

M. Moldaschl

Institut für Sozialwissenschaftliche Forschung e.V.,
Jakob-Klar-Strasse 9,
8000 München 40, Germany

E. Moritz

Wissenschaftlicher Assistent der Universität Hannover,
Institut für Soziologie
(Correspondence: Clemensstrasse 2, 8000 München 40, Germany)

R. Schultz-Wild

Institut für Sozialwissenschaftliche Forschung e.V.,
Jakob-Klar-Strasse 9,
8000 München 40, Germany

G. Spur

Institut für Werkzeugmaschinen und Fertigungstechnik,
Technische Universität Berlin,
Pascalstrasse 8–9, 1000 Berlin 10, Germany

J. Stahre

Department of Production Engineering,
Chalmers University of Technology,
S-412 96 Göteborg, Sweden

F. Zurlino

Fraunhofer-IPK,
Pascalstrasse 8–9,
1000 Berlin, 10, Germany

1 What is Human-Intelligence-Based Manufacturing?

Y. Ito

1.1 Introduction

During the last decade (from the late 1970s to the beginning of the 1990s), manufacturing systems have significantly changed. Their function and performance have altered in a period of rapid development, mainly as a result of the application of the techniques of computer science. The new systems are termed either flexible manufacturing systems (FMS) or computer-integrated manufacturing (CIM). The advance of flexible manufacturing and its related technologies has been accompanied by a number of related developments, such as flexible manufacturing of a global network type, CIM incorporating personnel functions, skill-based manufacturing, and compact flexible manufacturing cells (FMC). These all have notable influences on the system configuration and operational technologies of flexible manufacturing: perhaps as a result, there is some confusion about their terminology. A number of differing nomenclatures are used (such as FMC, FMS, flexible or factory automation (FA), CIM and enterprise automation (EA), for the same fundamental technology.

The most serious consequence of this trend is that the core definition of an FMS is no longer appropriate. Figure 1.1 shows schematically the definition of an FMS proposed by Weck [1]. The key to this is concurrence of mutually related processing of materials and information at a required station, island or cell. This generally leads to a marked reduction in idle time while waiting for either information or material. In flexible manufacturing of a global network type, the definition is inappropriate, since material cannot be so readily transferred as information. A new and consistent terminology is required to describe present and foreseeable flexible manufacturing technologies. The term "flexible computer-integrated manufacturing structure" (FCIMS) has been proposed [2, 3] to describe systems with an hierarchical structure; this encompasses systems ranging from FMC to flexible manufacturing of a global network type. Currently, FCIMS is accepted mainly to refer to CIM, FMS and information networks, as shown in Fig. 1.2. In this case, CIM is the integrated software function that processes product related (i.e. technical), production management and also management information. FMS is regarded as the hardware used at the factory floor level, interfacing to the lowest level of CIM. Additionally,

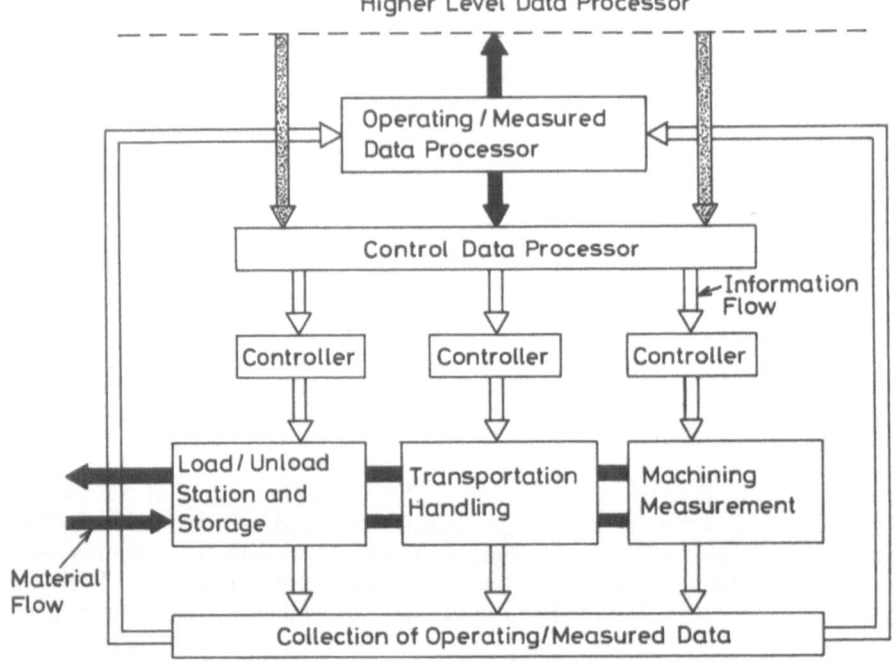

Fig. 1.1. Definition of FMS proposed by Weck.

FMS is, in principle, modular, using FMCs of various types, and compact FMCs as the basic modules [4].

Figure 1.2 elaborates the possibilities of using human intelligence-related functions to facilitate the development of a future intelligent FCIMS of global network type. Section 1.3 provides further details.

1.1.1   History and Future Evolution Trends of FCIMS

The history of FCIMS has resulted in confusion over its terminology: Fig. 1.3 outlines its history and possible future evolution. At their outset in the early 1970s, computer-aided design (CAD), computer-aided manufacturing (CAM) and FMS were developed independently of one another. FMS progressed to versatile production systems such as FMC and large scale FMC, FMS and large scale FMS, FMS integrated with CAD and flexible transfer lines (FTL). CAD, on the other hand, developed into CIM by incorporating CAM, computer-aided process planning (CAPP), material requirement (or resource) planning (MRP) and information processing for enterprise management. FMS and CIM are now being progressively integrated into systematically designed tools to achieve rationalization with production activity. The result is to produce FCIMS.

Figure 1.3 shows that FCIMS may evolve to assist in "one of a kind production" and production to match a customer's needs, so that the production process becomes much more responsive [5]. Long term predictions

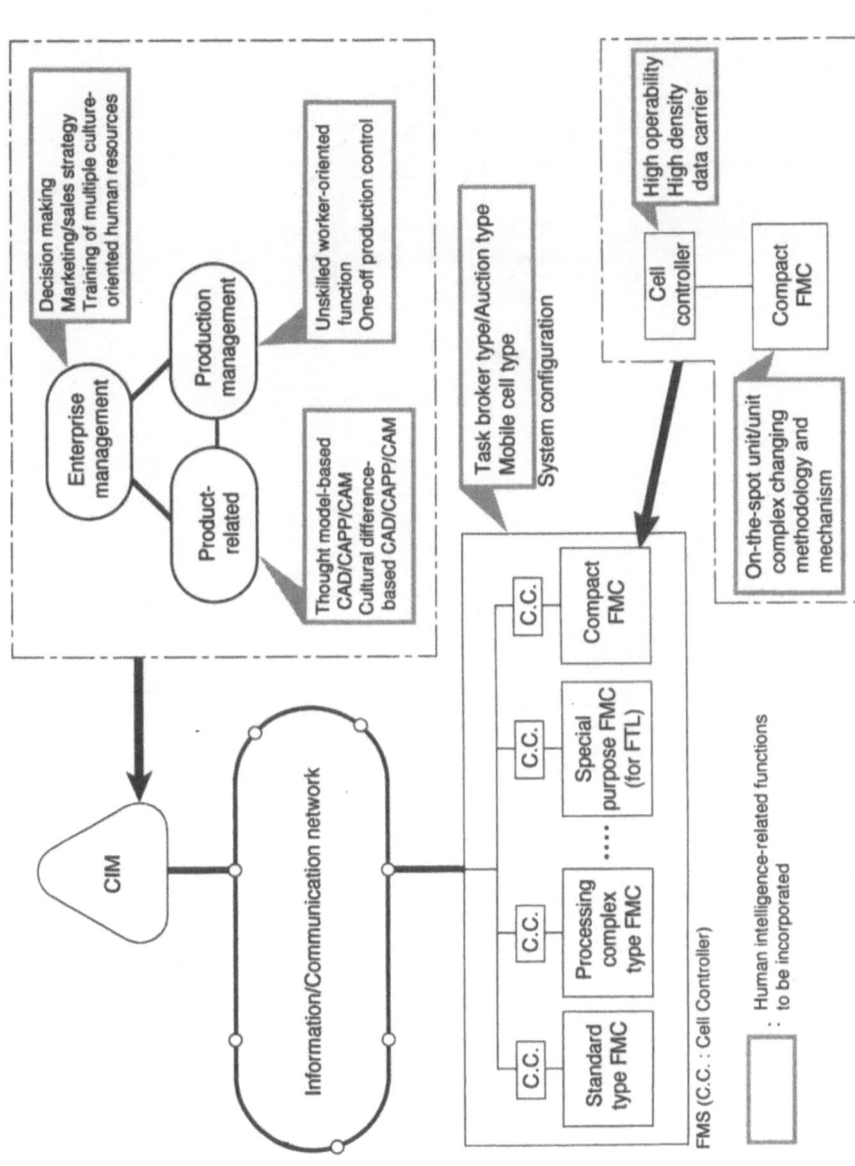

Fig. 1.2. Concept of FCIMS.

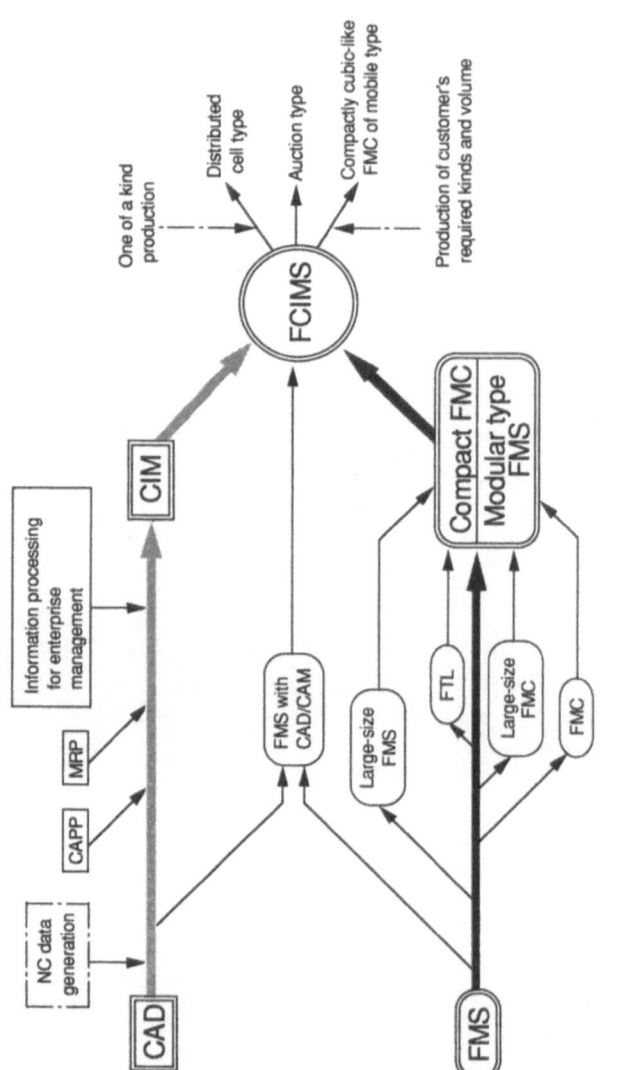

Fig. 1.3. History of FCIMS.

of the likely environment in the year 2000 were made between 1980 and 1985: these predictions may now be reviewed in the light of subsequent developments [6]. Table 1.1 describes the key factors (words) describing expected future production environments. In the table, a circle denotes a relation between a proposer and a model. Figure 1.4 illustrates some possible system configurations derived from the models in Table 1.1.

The main interest then is to predict the likely technical trends in production environment in the years ahead and to ascertain how those might affect and be affected by the economic and social climate. FCIMS is becoming more mature and has been evaluated against these criteria in trials of its prototypes [3]. Figure 1.5 shows the maturity trends for FCIMS: note the human-oriented type, since this points out the continuing involvement of humans in the production environment [7]. Whilst machining processes are believed to become fully automated, the assembly process is likely to require significant use of labour into the future. Even in machining, there is a requirement for experienced maintenance staff to supervise sophisticated factory floor systems. Human-oriented FCIMS may be interpreted as a major component of "human-intelligence-based manufacturing" (HIBM). FCIMS of global network type should be taken alongside human-intelligence-based manufacturing, providing that the system design takes account of the "culture of manufacturing", as described later. For instance, a system design should take account of its cultural context.

Now considering the core technologies for FCIMS in the future (reviewed by Ito [7]), the following may be described in relation to Fig. 1.2.

1. Compactly cubic-like FMC. This may be considered as a core module of the future FMS, and can be classified into (a) standard, (b) multiple-processing function-integrated and (c) system function-integrated types. Figure 1.6 illustrates a current compact FMC of system function-integrated type. Figure 1.7 shows a concept of a mobile compact FMC and the system configured by it [2]. The concept was recently proposed to extend the idea to establish highly flexible manufacturing in the long term using modular design.

2. Multiple-function integrated transducer and sensor fusion for in-process measurement.

3. CIM of artificial intelligence (AI) type, which can also manage information related to manufacturing culture and human deep knowledge.

In the near future, FCIMS could be of a distributed cell type so that it might be significantly assisted by the data tag (carrier: an IC card or tip) – that is a core technology which simultaneously transports materials and information. The arrival of this new technology stimulates the need for further study of cell controllers so that they should be able to operate with autonomously distributed information processing. Figure 1.8 shows the concept of such cell controllers [7]. In this case, a workpiece to be processed is transferred along with data. Each cell can decide whether the workpiece is acceptable by comparing the data recorded with the cell's configuration. The system concept is similar to that of an auction, or task broker, type which aims to enhance the system flexibility by emphasizing its software-oriented function.

Table 1.1. Key words for forecasting future FCIMS

Proposer	Key word	Factory location				Manufacturing mode				Factory/system configuration		
		To be close to market	Decentralized production structure (Strategical distribution of cells)	Global network	Use of space/ underground/ sea	Lot size of one	Large-volume production	Operation of 24 hrs/day, everyday	Totally computer integrated/ flexibly-automated	Compact factory	Modular structure	High reliability (planned backup/ redundancy)
USA	Slautterback	O	O			O		O	O (Unmanned)	O	O	O
USA	Swyt					O (Custom quantity)	O		O	O	O	
USA	Harvey					O			O			
Germany	Spur		O		O				O	O	O	O
U.K.	Connolly		O			O		O	O	O		
Japan	IROFA	O	O	O	O	O				O	O	

Proposer	Major hardware				Major software			
	Intelligent system component	Advanced robotic-based	Less need for warehouse	Communication network	Integrated information network system	CIM (Including AI type)	Simulation system	CUMS/CAS
Slautterback		O	O	O	O	O	O	
Swyt	O	O			O	O		
Harvey			O	O		O		O
Spur	O (To replace machine tool)	O		O	O (Centralized management)	O	O	O
Connolly		O		O	O (Enhancement of marketing information)	O	O	O
IROFA		O		O	O	O		O

O : Pointed out by each proposer   Note)   CUMS : Customer Requirement Managing System
                                                   CAS   : Computer Aided Service Planning

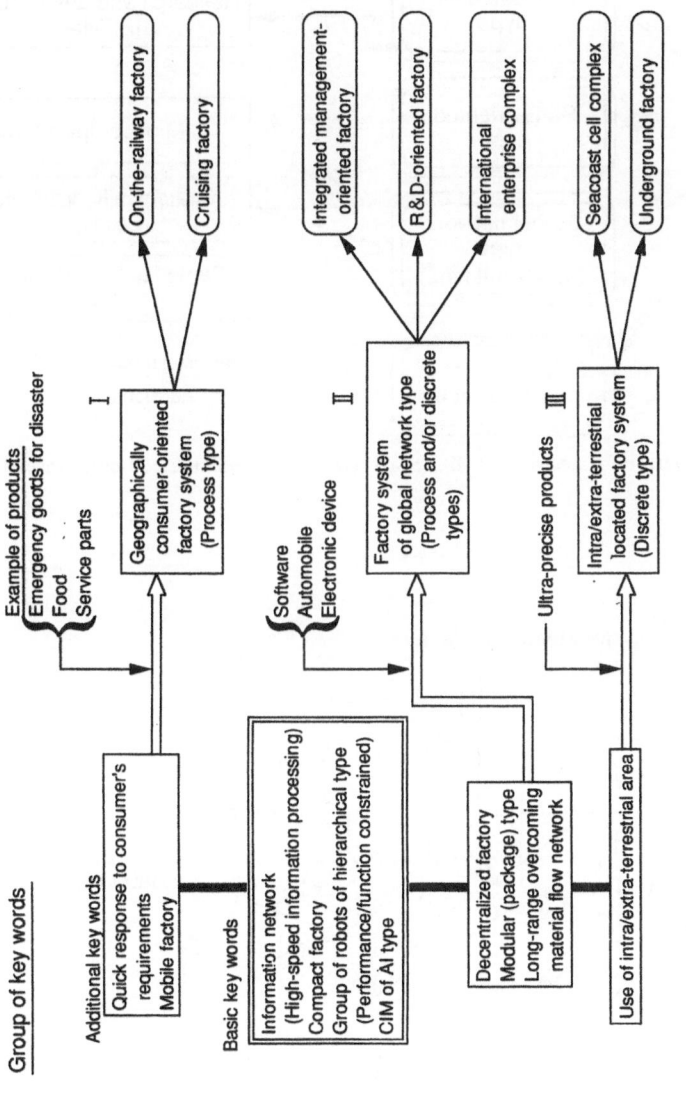

Fig. 1.4. Basic system configurations of future FCIMS and their variants.

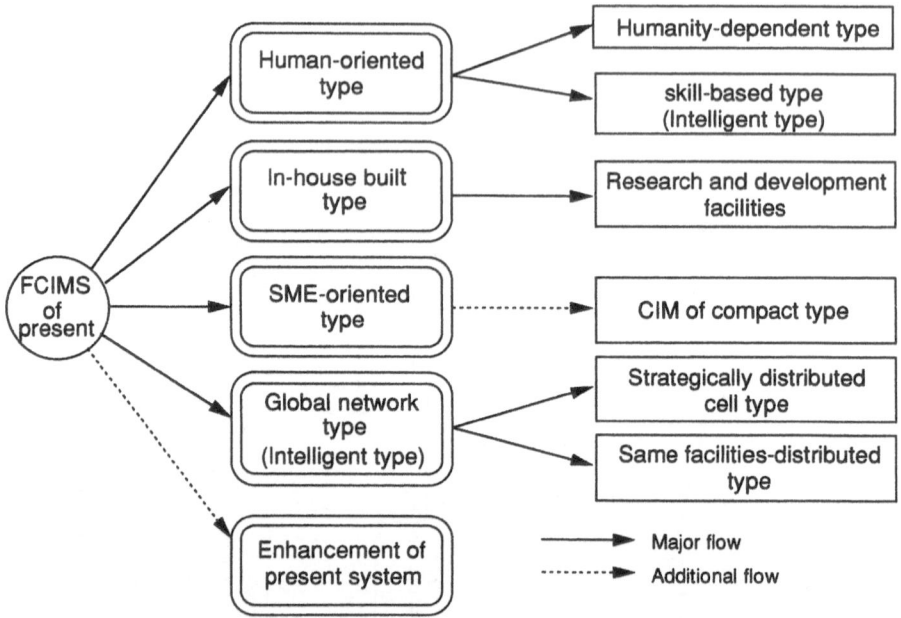

Fig. 1.5. Maturity states of FCIMS. (SME: small and medium-sized enterprises.)

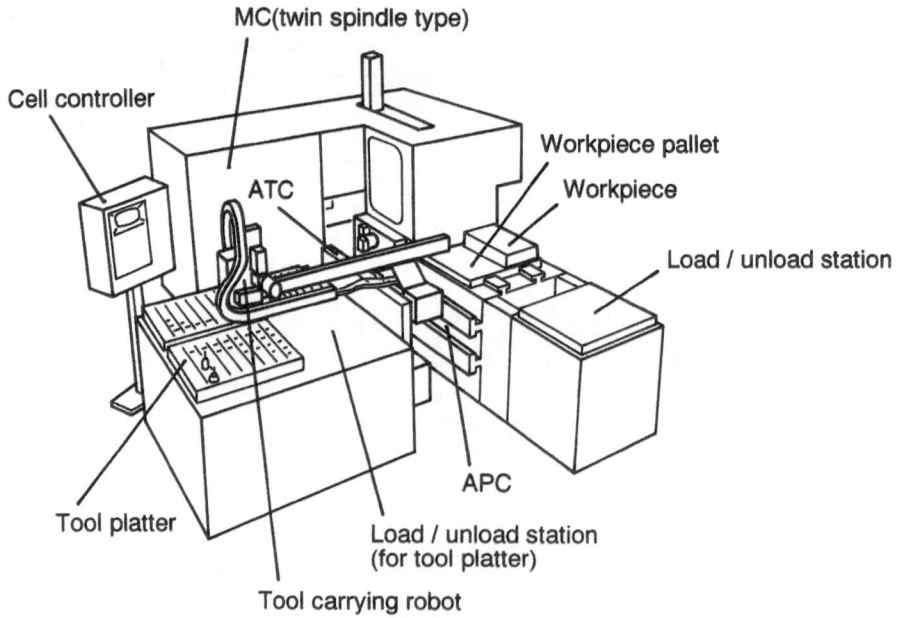

Fig. 1.6. Example of compact FMC (MPM Co. Type: AUTOMAX I).

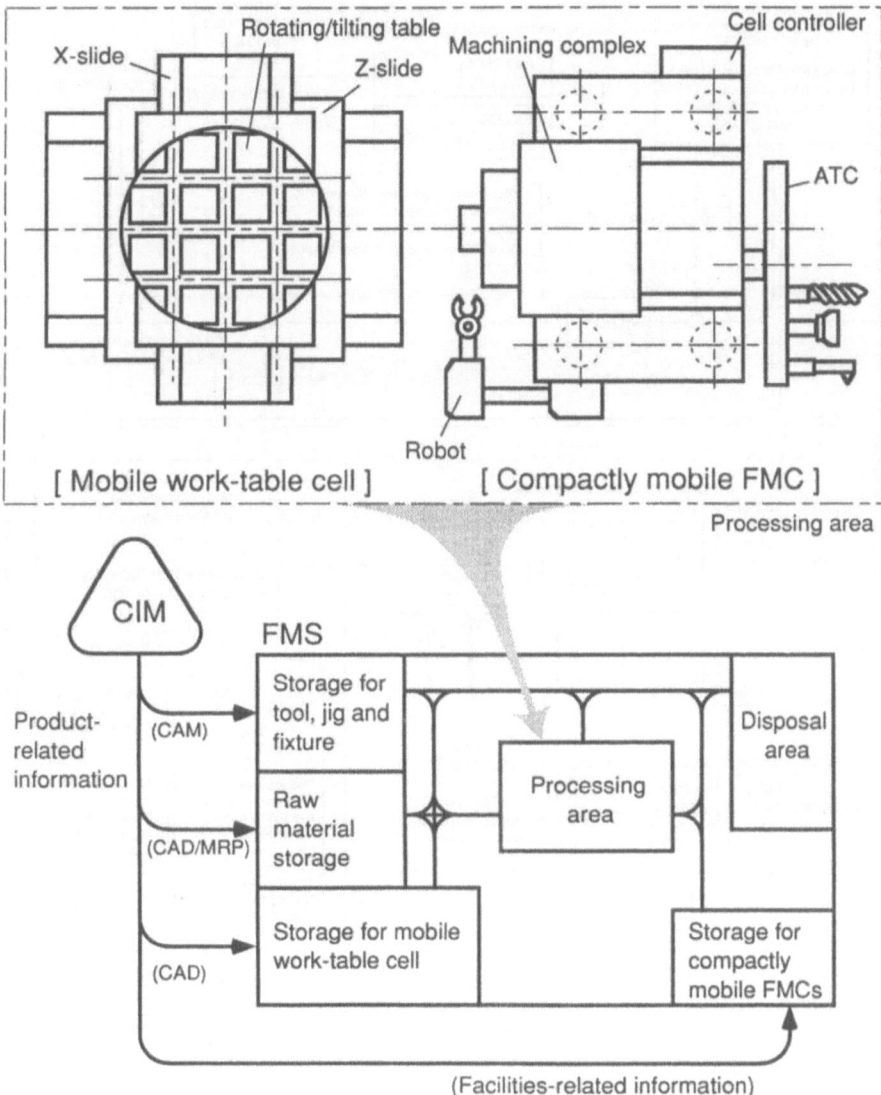

Fig. 1.7. FMS configured by a group of compactly cubic FMCs of mobile type.

1.1.2 Necessity and Potential of Human-Intelligence-Based Manufacturing

We may consider the following facts:

The difficulty of establishing fully unattended systems, except FMS, for the machine tool-based industry. As was mentioned above, the important roles for human involvement in future production environments have been

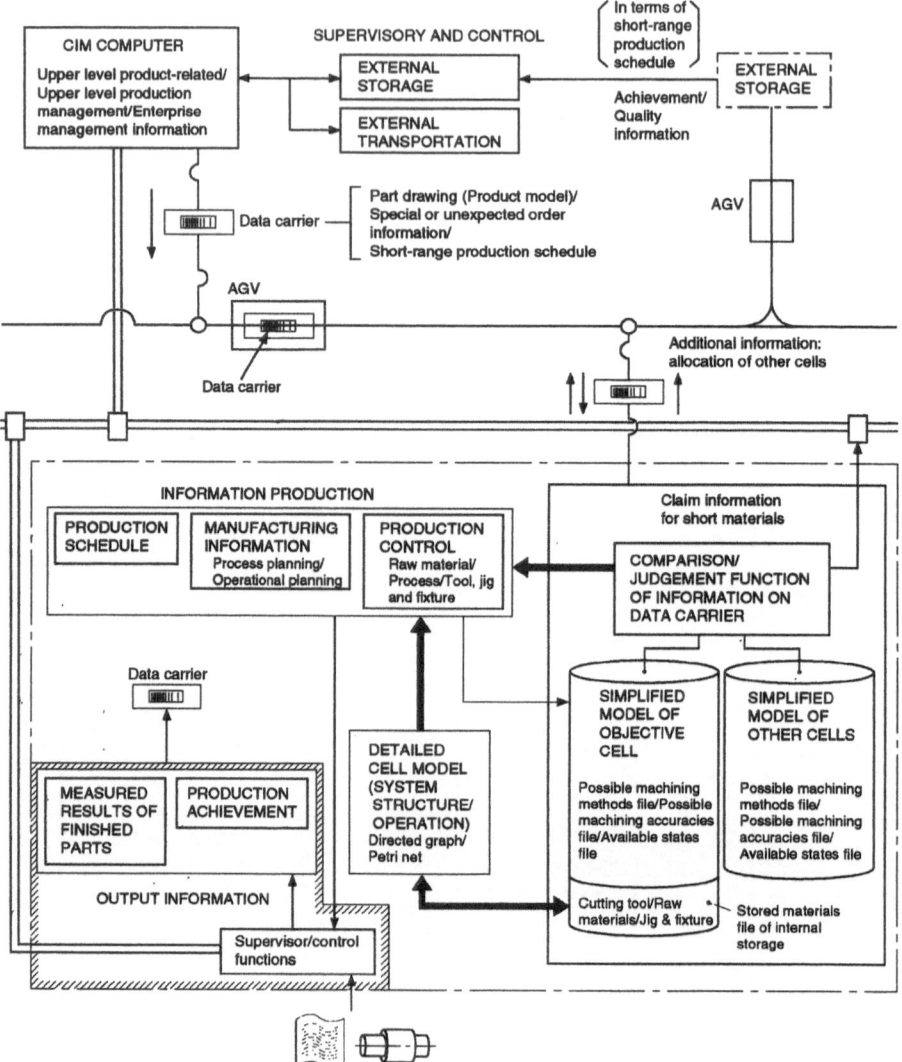

Fig. 1.8. Concept of intelligent cell controller compatible with future production environment.

described by most reports relating to forecasting the production environment in the year 2000. Table 1.2 summarizes the roles of the enterprise manager and worker [6]. In flexible assembly systems, as shown in Fig. 1.9, it is recommended that a worker should be located in the final assembly line to handle any soft materials (such as rubber tube, electric cable and wire). This is because the automated facilities and robots lack the dexterity of a human.

The requirement to have an experienced worker effectively and efficiently to operate the system. The operational efficiency of flexible manufacturing

Table 1.2. Major human roles in future FCIMS

Key word / Proposer	Enterprise management			System management/control		Factory floor			
	Strategical market prediction	Planning of integration/ innovation strategy	Education/training of multiple-functional man-power	Supervision/operation of CIM (by a few highly trained specialists)	R&D of utilization technology	Parts setting in assembly/ machine setting	Inspection	Maintenance	Elimination of defects/quick repairs
Slautterback	O		O		O	O		O	
Swyt			O						
Connolly	O		O			O	O	O	
Spur		O	O	O				O	O
JEIDA						O	O	O	

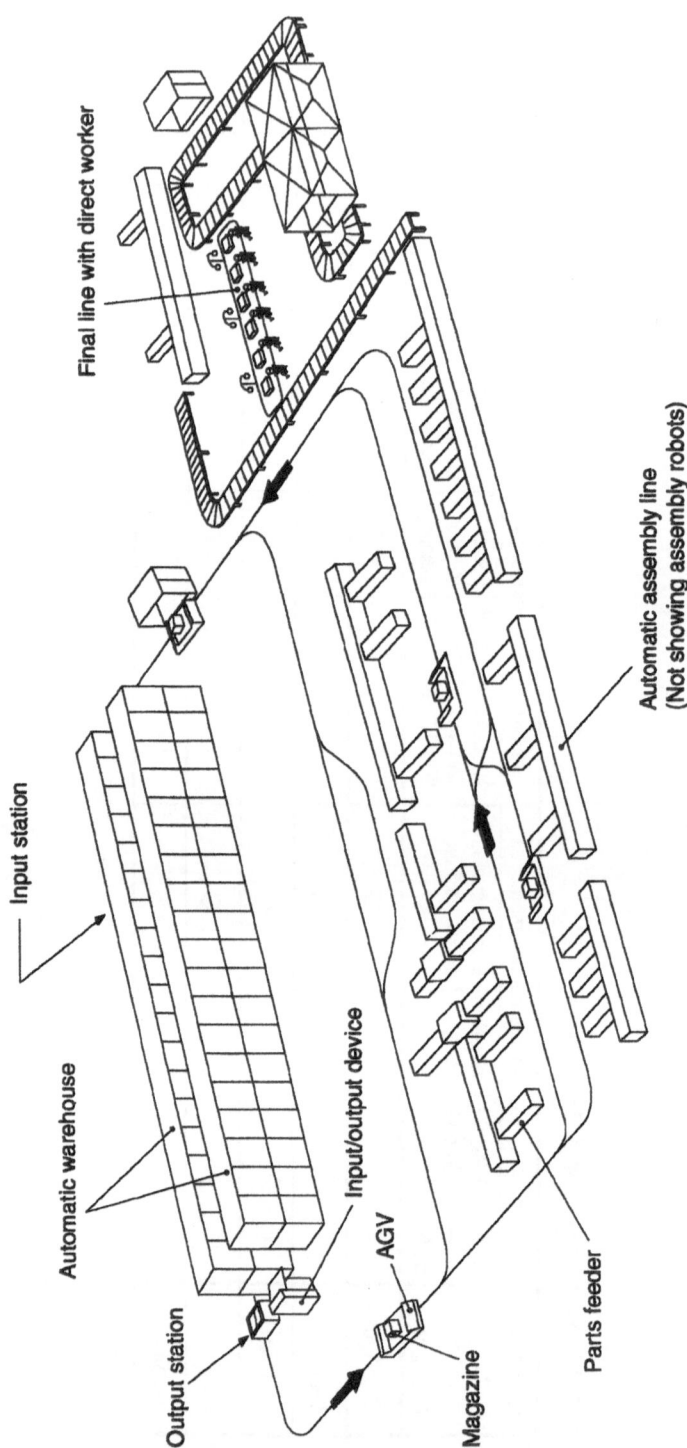

Fig. 1.9. Flexible assembly system for dot matrix printer (Tohoku Oki Electric Co., Japan).

systems depends largely on the skill and experience of a worker. For instance, a flexible machining system is designed to work effectively, as long as swarf removal and work setting are satisfactorily carried out. Generally it is difficult to replace the skills of a worker with automated facilities. These possibly could be overtaken by expert systems, although there is still concern about the limited applicability of such systems to those tasks.

The need to introduce the flair, thought processes, skill, experienced knowledge and so on to the computer environment in the production of highly value-added goods. In the already industrialized countries, one of the next objectives is to establish the manufacturing technology required to produce highly value-added goods. The core technologies in this respect appear to be those which seek to emulate the flair, thought processes, experience and skill of the mature engineer and technician. They also must take account of people's perceptions. This requires the establishment of new CIM based on another concept which is simultaneously applicable to FCIMS of global network type explained later. In contrast, present CIM of AI type is generally aimed at rationally integrating production, production management and enterprise management information. The essential features of an extended CIM of AI type are shown in Fig. 1.10, which also highlights the differences between the CIMs outlined above.

Consideration of the culture of manufacturing in FCIMS of global network type. It is widely recognized that fashions of production spread globally. These reinforce both international co-operation and competition. The core technology here is also a new CIM whose most important functions relate to the "culture of manufacturing". For example, software or information

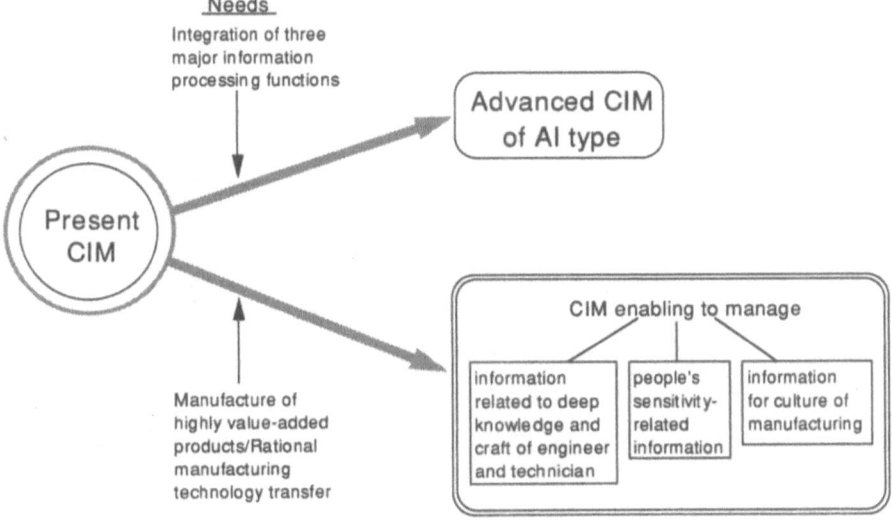

Fig. 1.10. Advanced CIMs of two different types.

processing for supporting decision making should take account of the culture where it is to operate, geographical situation, historical background and so on.

As described above, HIBM is of central importance to the design of future production systems. This will continue to be the case even as the degree of automation increases in future. In establishing HIBM, CIM based on these new concepts will diverge markedly from the present CIMs of AI type, although both strategies will rely on AI technology. Meanwhile "Anthropocentric computer-integrated manufacture" has been proposed, especially by European researchers. For our purposes, this is considered to be encompassed by HIBM.

1.2   Representative Human-Intelligence-Based Manufacturing Systems and Their Characteristics

Before considering the details of HIBM, we should review its position in relation to the developments in FCIMS which are shown in Fig. 1.11 [2]. This shows that HIBM has two facets: one is based on the extension of current FCIMS, whereas the other aims at the configuration of the mainstream of future production structures. In each case, as we have observed, the trend is to emphasize the importance of human-oriented system functions, whilst maintaining productivity. Owing to the variety of interpretations, the precise definition of the term HIBM has not yet been agreed.

A number of HIBM systems have now been realized. Those of which we are aware are in their various stages of development, prototype and

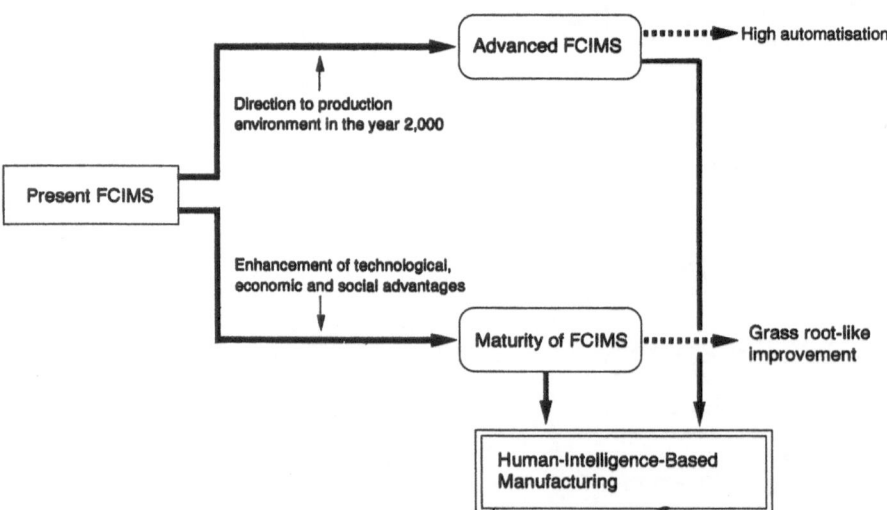

Fig. 1.11. A developing map from FCIMS to HIBM.

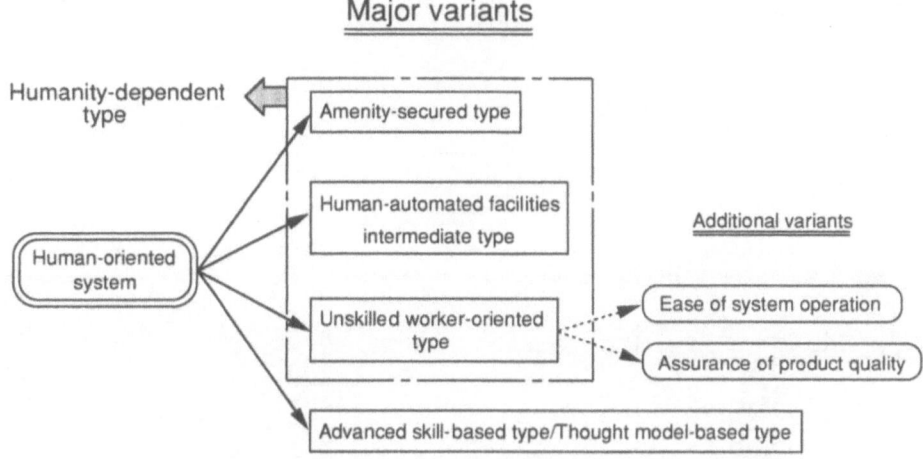

Fig. 1.12. Various evolution trends in human-oriented FCIMS.

application to production facility are summarized in Fig. 1.12 [3]. Of those listed, both the amenity-secured, human-automated facilities of intermediate type, and of the unskilled worker-oriented type are in practical use: the number of systems remains small. The following sections clarify our understanding of HIBM: further details of each type are presented in the succeeding chapters.

1.2.1 Human-Amenity-Secured Type

The earliest references to HIBM relate to the human-amenity-secured type. The trial at the Kalmar plant of the Swedish Volvo Company is one of the most famous examples. It employs units of a hexagonal shape and has the following features:

1. Car assembly is carried out by teams in each hexagonal unit.
2. The favourable effects of sunshine are admitted by making the walls of the hexagonal units of glass.
3. Each unit has a "green zone" for refreshment to add to the warm feelings brought about by the sunshine.

Figure 1.13 illustrates the layout of Volvo's Uddevalla plant which is one of the most recent in that company, and may be regarded as a successor to the Kalmar plant. Here, a team of about five people has the responsibility for the whole assembly process of a car, including its quality, and maintaining productivity. The team is organized with regard to both its age balance and skill maturity. Additionally, the plant has many green areas for refreshment and a relaxing atmosphere, resulting in stress-free assembly.

As was outlined above, the Volvo Company has endeavoured to establish complete manufacturing systems of the amenity-secured type. We should, however, be aware of a point of some controversy: the discrepancy between increases in productivity and enhancements to human amenity. No Japanese

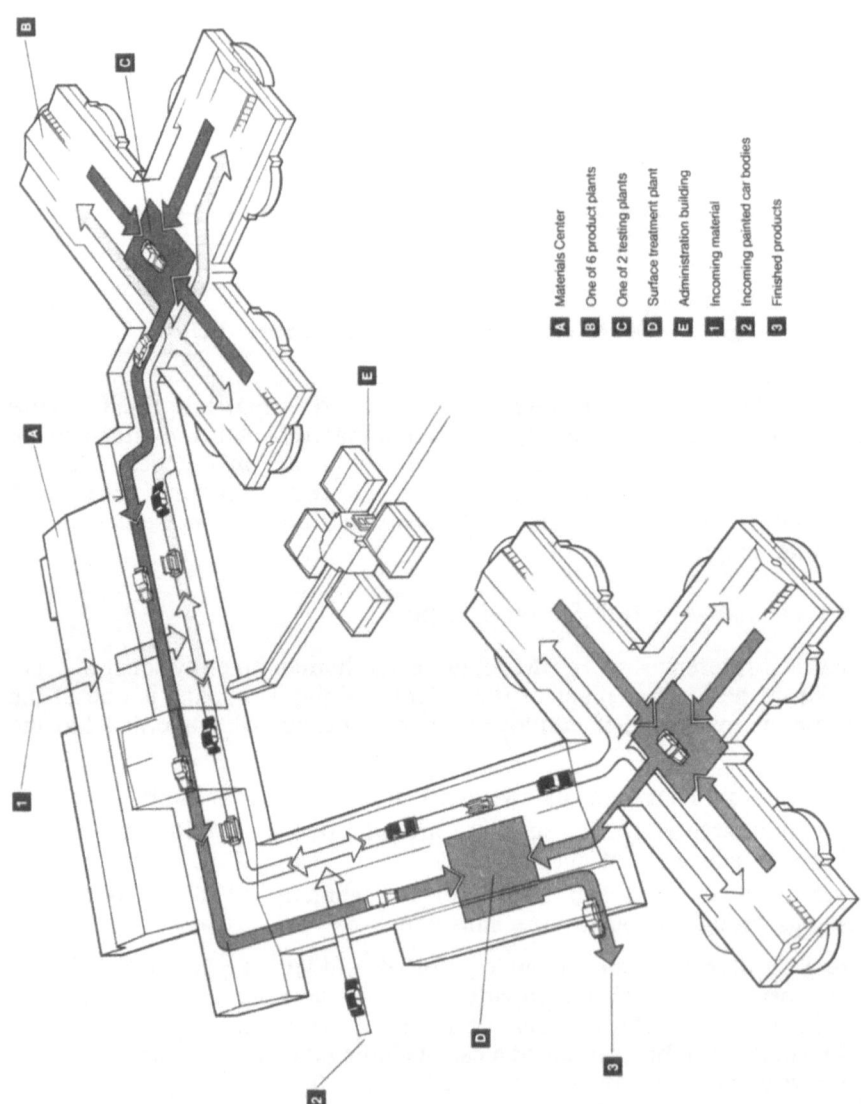

Fig. 1.13. Basic layout of Uddevalla Plant of Volvo Car Co. (by courtesy of Volvo Co.).

A Materials Center
B One of 6 product plants
C One of 2 testing plants
D Surface treatment plant
E Administration building
1 Incoming material
2 Incoming painted car bodies
3 Finished products

manufacturers are considering emulating Volvo's methods, since the methods are considered to reduce productivity. However, some engineers have studied Volvo's methods from the viewpoints of system design and the interface between technology and sociology. These views lead us to propose a new academic field – "socio-mechanical engineering". Volvo's systems were designed as a result of co-operation between engineers and sociologists.

1.2.2 Human-Automated Facilities of Intermediate Type

Human supervision will continue to be a very important factor even as automation and unattended operation increase at the level of machine, station or cell. Any new developments should therefore seek to optimize the use of both human and automaton. The choices will be made from the results of studies of both social science and technology. Since the flexible assembly systems used now in the electronics industry show considerable dexterity, they are close to being classified as human-automated facilities of intermediate type.

Such manufacturing systems consist primarily of workers, assembly robots and a transfer line. It is of prime importance therefore to consider the characteristics of each before allocating their respective tasks. A worker might handle wiring and the assembly of parts made from soft materials, but the robot would undertake simple assembly, such as fastening screws, greasing and insertion. Table 1.3 illustrates some of the systems so far implemented by the Sony Company and its subsidiaries. Typically in these examples, a robot replaces one or two workers. Such robots should cost less than 1 million yen.

1.2.3 Unskilled Work-Oriented Type

This type was first characterized as HIBM in a Japanese context. The Japanese economy has long faced both labour and skills shortages. The shortfall has accelerated the development of unskilled worker-oriented systems with sufficient configurability and function for simple operation and high product quality. These systems are applicable in developing economies where the simplification of operations may ameliorate the skill shortages. Increasing requirements in developing countries for flexible manufacturing

Table 1.3. Examples of human-automated facilities of intermediate type

Name of company	Objective product	Tasks performed by robot	Tasks performed by people	Kinds of robot
Sound Systems	Audio-visual devices	Thread tightening/ Greasing/Soldering/ Lead wire cutting	Work preparation	Co-ordinate, bench type
Sony Co.	Car CD player	Thread tightening	Work change/ Change of thread	Co-ordinate, bench type
		Thread tightening/ Greasing/Work transportation	Work preparation/ Final assembly	Co-ordinate, bench type/ Scara type

technologies have meant that the application of unskilled worker-oriented systems have given rise to international co-operation.

Perhaps surprisingly, highly automated production requires the skill of workers to ensure that it functions successfully and efficiently. For instance, metal cutting using an FMS needs skilled intervention to handle both swarf removal and to ensure that the tooling layout is correct. Part-time workers are unlikely to be sufficiently skilled, so the system must be provided with an adequate man–machine interface with visual indicators to ensure safe operation.

Figure 1.14 shows one such system that produces programmable controllers (PC). It is operated at the Mishima plant of the Omron Company. Even though the yield rate of the product is significantly affected by the organization of work preparation (picking up required parts and subassemblies and loading them into a bucket), a part-time worker can successfully manage with the assistance of technology for guidance.

1.2.4   Some New Approaches to HIBM

We now extend these arguments to new approaches to HIBM. One is the "culture of manufacturing" and another, "thought and skill models".

The former has become a serious problem with the advent of globalization in production for both large and medium- to small-scale enterprises. In other words, enterprises have been forced to diversify their production activity worldwide; however, FCIMS of global network type is not yet fully realized. Current production structures based on wide area organization networks have been implemented to a certain extent, but these do not constitute FCIMS of global network type. Table 1.4 provides a summary of some trials that aim to establish real FCIMS of global network type [3].

Our foremost problem is thus to clarify the main reasons why FCIMS of global network type has not yet spread. The culture of manufacturing is a powerful means to move this problem forward as it provides considerable knowledge of system design linked to the local circumstances where a

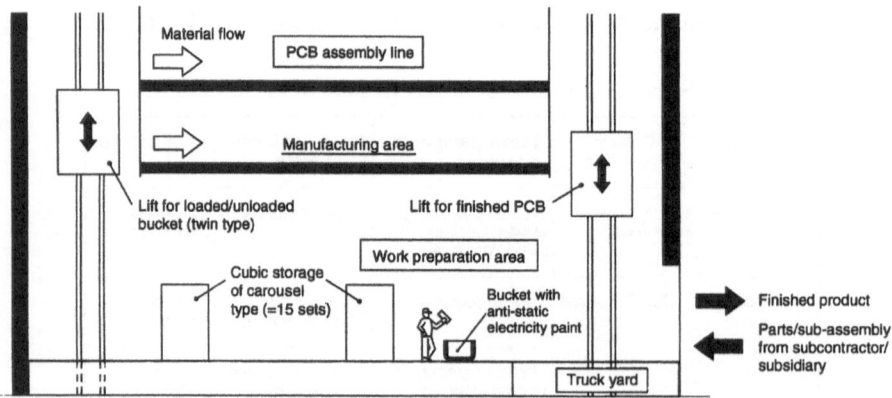

Fig. 1.14. FCIMS of unskilled worker-oriented type (Mishima Plant of Omron Co.).

Table 1.4. Trials of FCIMS of global network type

Name of organisation	System configuration	Objective product	Remarks
Kubota Iron Works (Japan)	Parent company-subsidiary	Petroleum/Diesel engines	JIT for part supply with VAN Collection of parts to be supplied with circulating lorry
Nihon Seiko Co. (Japan)	CIM network within an enterprise (Including executive's room-factory floor)	Rolling bearing	For production/enterprise management
Omron Co. (Japan)	CIM network within an enterprise	Programmable controller and others	For user's order/production management
MBB AG (FRG)	Plant to plant within an enterprise	Aeroplane	————
Austin Co./Lucas Co. (U K)	Between different enterprises	Electric/electronic car parts	————
Toride plant of Canon Co. (Japan)	Same facilities being decentrally located over the world	Camera/Copying machine	————
Sumitomo-Dunlop Co. (Japan)	Decentral location of different facilities compatible with each regional characteristic	Tyre	————

system might be installed and operated. Table 1.5 illustrates the factors that might be considered when designing FMS for use in Asian countries [2].

Table 1.5. Indigenous design specifications for Asian country-oriented flexible manufacturing systems

Country	Design specifications
Korea	Relatively cheap labour cost
Taiwan	Poor ability for large investment
P.R. of China	Sudden stop of electricity supply
Philippines	Immature machine tool industry/Sudden stop of electricity supply
Malaysia	Relatively high NC utilisation technology/Immature machine tool industry
Thailand	Outstanding utilisation technology for only transfer line

Meanwhile the industrialized nations seek to develop production structures for highly value-added goods that are compatible with their circumstances. It is still unclear which products these may be in the future. Examples of these products as proposed by the International Robotics and Factory Automation Centre (IROFA)[1] of Japan are:

---

[1] IROFA is a Semi National Body Organization under the control of MITI.

- Information-related devices
- Prosthetic devices
- Individual-oriented consumer goods
- Security-related devices
- Cars/super-compact mobile vehicles
- Construction machines/agricultural machines
- Space-related devices/ordnance
- Machine tools
- Production systems/system designer (man-power)
- New hardware/high operability devices

It is believed that most future products are likely to be human sensitivity-related goods, prosthetic devices, aerospace-related and aesthetic (artifact-like) products. The controversy lies in deriving an appropriate manufacturing structure for these highly value-added artefacts. Proposed design specifications for these production systems are shown in Table 1.6. Prosthetic devices could be produced by such a system which incorporates functions to manage personalization of products. This leads to the idea that "skill-based manufacturing" could respond to the production of a highly value-added product.

In this case, we must remain aware that skill-based manufacturing can evolve in two directions. Although we use one terminology, skill-based manufacturing embraces two different system concepts. One puts its main stress on the human factors, such as skill, know-how, job enhancement and enrichment, thereby increasing the operational efficiency of FMS and maintaining the worker's own level of skill. Productivity in this model overrides either humanity or human amenity, although they are paid some attention. People involved in industrial sociology are thus especially interested in the organization of skill-based manufacturing [8].

The other concept aims at the production of highly value-added goods. This model should be called "thought model-based manufacturing system" (THOMAS) [2]. People are expected to play a central role in such manufacturing. The core technologies being applied analyse flair, intuition and the thought processes of the engineer and designer, crafts skills and personal sensitivities.

1.3 General Trend of Academic Research Activities on Culture of Manufacturing and Thought Model-Based Manufacturing

Although much work remains to be undertaken, HIBM and some related research areas have been addressed in fields known as "culture of manufacturing" and "thought model-based manufacturing". Table 1.7 clarifies summaries of these research topics [2]. This table emphasizes the approach which seeks to amalgamate the study of automation technology with research into human intelligence. This work points to a new academic discipline "socio-mechanical engineering". This will highlight the growing proximity of engineering with sociology as well as economics.

Table 1.6. Design specification for system to produce prosthetic devices

Outlines of system / Products	Factory location			Production mode		System configuration	
	Close to market	Strategically distributed allocation	Intraterrestrial area	One-off production (Individual difference compatible production)	24 hrs continuous operation	Computer-integrated/ flexibly automated	Cube/Compact
Prosthetic devices							

Major system components				Others
Diagnosis/CAD/CAM integrated system	Information communication network	Customer's service information system	Renewal/repair parts centre (Quick delivery system)	Usage of new material processing/micro-mechatronics

Table 1.7. Examples of imperative future research subjects on THOMAS

	Research subjects
Culture of manufacturing	Design methodology for Asian Country-oriented type
	Conversion methodology from social-related information to system design specification
	Technology classification into commonly available and indigenous ones
	Cell controller of advanced AI type
Thought model-based manufacturing systems	Quantification of hackneyed phrase used in design work, e.g. good designed drawing giving us stable and comfortable feeling
	Amalgamation of engineer's intuition with CAPP of neural network type
	Design methodology for individual difference-oriented product
	Colour glossiness monitoring system
	Cell controller of thought model-based type

1.3.1 Research Subjects on Culture of Manufacturing

As Table 1.7 demonstrates, the topics are diverse, as they range from "design and utilization strategy of CNC machines among USA, Japan and Germany" to "influences of family, social and enterprise factors on production systems". This is due to the multi-faceted nature of the culture of manufacturing itself. The research activities have been carried out by workers in different fields. For instance, sociologists carried out the related research into "industrial culture". It is now beneficial to amalgamate these diverse topics.

Whilst the importance of any work will depend on a personal perspective, some pertinent points are summarized here to deepen our understanding of the "culture of manufacturing". Details are given in Chaps. 8 and 9.

Table 1.8 lists findings of some characteristics of the Japanese culture of manufacturing. These were the results of an on-the-spot investigation by Kolar [9]. In this case he and his colleagues investigated the effects of Japanese culture on production activity. They emphasize three factors related to family and society, education, and enterprise organization.

By contrast, Fig. 1.15 compares the methods of machine tool design in Japan and Germany. It may be seen that there are notable differences in the procedures stressed in the design flow. The Japanese, for instance, endeavour to obtain correct market information, whereas the German designer is more concerned with the generative design concept [10]. This implies that a technology consists of two areas: one is commonly applicable to all countries, and another is specific to each. As this example demonstrates, the technologies related to the culture of manufacturing could be very effective in ameliorating the conduct of international co-operation. Sometimes they might reinforce international competitiveness.

To further our understanding, Fig. 1.16 and Table 1.9 show typical design

Table 1.8. Characteristic aspects in Japanese culture of manufacturing (Kolar et al. [9])

Items		Remarkable findings
Family & social factors		Concept of inner and outer world
		Hierarchy according to age
		"Education Mania"
		"Filial piety"
Academic factors		Large population of mechanical engineering students from non-elite schools going into manufacturing
		Large population of engineering graduates having Ms.-degree, but not having MBA
		Most of top managers have, at least, an undergraduate degree in engineering
Industrial factors		Business is run to employ people
	Recruitment	Human resources : university graduates successfully completed many years of competitive schooling
		People able to change professional fields easily
		Often engineering-trained personnel manager
	Training & development	Requirement of shop floor working for every new engineer
		Heavy investment in training & development programme
	Workplace	Incorporation of various aspects of Japanese culture into workplace
	Promotion and personnel appraisal	Equal opportunity for the same raise per each year at roughly the same age, but promotion with different responsibility
	Compensation	For instance, rent of company-owned housing
	Corporate relationships	Bonus system
		Underlying society and culture among first, second and third tier companies
	Labour relations	Enterprise union
		Even senior director was once an officer in the union
	Small group activities	Similarity between "Circle" and "Family"
	Non-regular employees	Successful employment of part time workers
		Employment of foreign workers for most stressful type of manufacturing

methods for mechatronics and related products in Japan, along with their "superiority factor" [11]. They demonstrate why Japanese products in these areas are strongly competitive in world markets, but also highlight their weak points.

1.3.2 Research Subjects on Thought Model-Based Manufacturing

Figure 1.17 illustrates a tree structure that contains the research subject in thought model-based manufacturing and skill-based manufacturing. It shows some ideas on the nature of thought model-based manufacturing and its

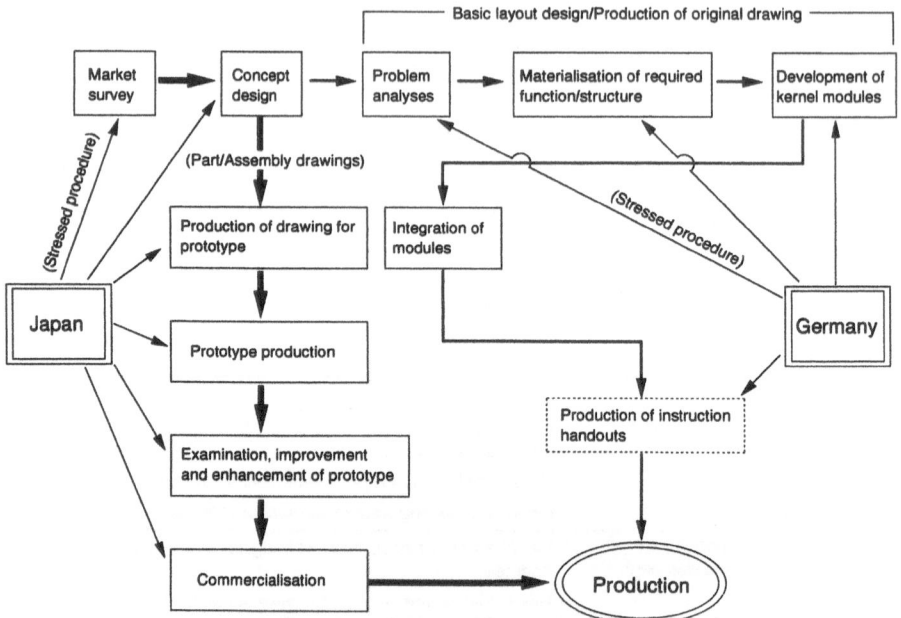

Fig. 1.15. Comparison between German and Japanese approaches in machine tool design.

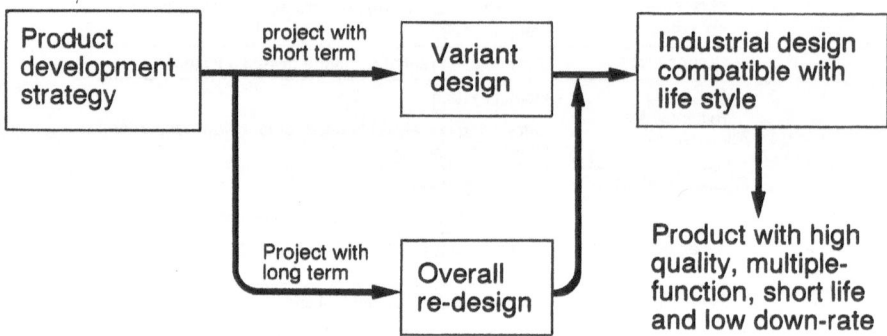

Fig. 1.16. Design flow for mechatronics and related products in Japan (Buur [11]).

likely evolution. The technology of thought model-based manufacturing is contrasted with that of traditional skill-based manufacturing.

Some pioneering research has now been carried out to unveil the flair and thought processes of an experienced process engineer [12, 13], and the decision-making processes of a mature designer [14]. The objective is to transfer such uncertainty-related information processing to the computer environment (details are given in Chap. 7). A designer's sensitivity has also been tested. The designer may evaluate the result of a design: "A well designed product or part has a well balanced shape and dimensions and provides us with a comfortable atmosphere". The validity of this expression

Table 1.9. Superiority factors in Japanese methods of mechatronics design (Buur [11])

Product development strategy	Emphases on enlargement of market share
	Determination of market price to arouse demand compatible with present production capability
	Special attention to user's famous brand worship
	Determination of product life based on user's demand
	Short product life —— *Ex.* Consumer supplies : 3 months ~ 2 years
	Product with multiple-function
	Excellent ability for information collection and analysis of competitor's new product
	Manufacture of the same kind of product by many enterprises
Variant (version-up) design	Continuance of grass root-like improvement
	Incorporation of the newest technology even in partial improvement to arouse demand and to enhance competitiveness

is evaluated by examining the impressions of other people who assessed a two-dimensional diagram [7]. The conclusions of the analysis are:

1. Personal sensitivities may be unveiled using "factor analysis", and classified by the static and dynamic evaluation values.
2. A person's sensitivity is believed to correlate closely with the dynamic value, which can be objectively assessed by the attribute values in the diagram viewed.
3. The dynamic value depends strongly on personal circumstances, and their cultural and historical background.

It will be necessary to initiate some new research facilities on "thought model-based" manufacturing to meet the new demands being placed on it. Research facilities[2] recently have been installed at Tokyo Institute of Technology to conduct research activities on "intelligent machines" (Fig. 1.18). This includes a group of FMCs of intelligent type [3, 7] which are being used to investigate the following subjects.

1. Analyses of the flair and thought processes of an experienced process engineer. Attempts to transplant this knowledge to the computer environment, e.g. CAPP of flair type.
2. Analysis of the decision making processes in the basic layout design of machine tools, and the evaluation of the "thinking block" used by the designer [14].
3. Development of a cell controller of the intelligent type such as shown in Fig. 1.8.
4. Development of in-process sensing technology for ambiguous quantities to replace the five human senses.

---

[2] The official name and its acronym are the "Advanced (Anthropocentric) Intelligent Machine Complex", AIMAC.

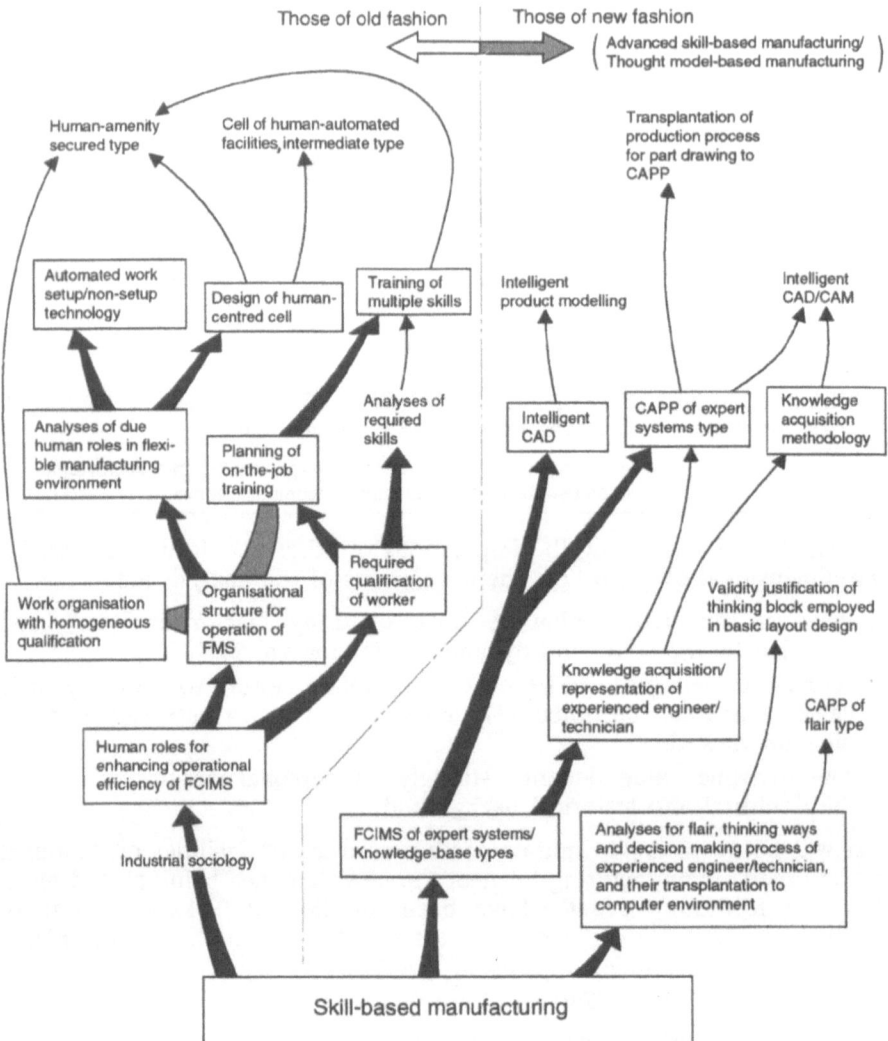

Fig. 1.17. Tree structure of research subjects in skill-based manufacturing.

1.4 Concluding Remarks

The advance of FCIMS and its related technologies has enabled us to characterize a variety of system structures, some of which are in active use. As it has developed, FCIMS has gradually incorporated some ideas involving the interaction of humans with both design and manufacturing processes. The systems may now be termed "human-intelligence-based manufacturing", which sounds like a very new field, and its novelty is largely true. From the viewpoints of both manufacturing requirements, and applicable areas, human-intelligence-based manufacturing is a completely new discipline, with

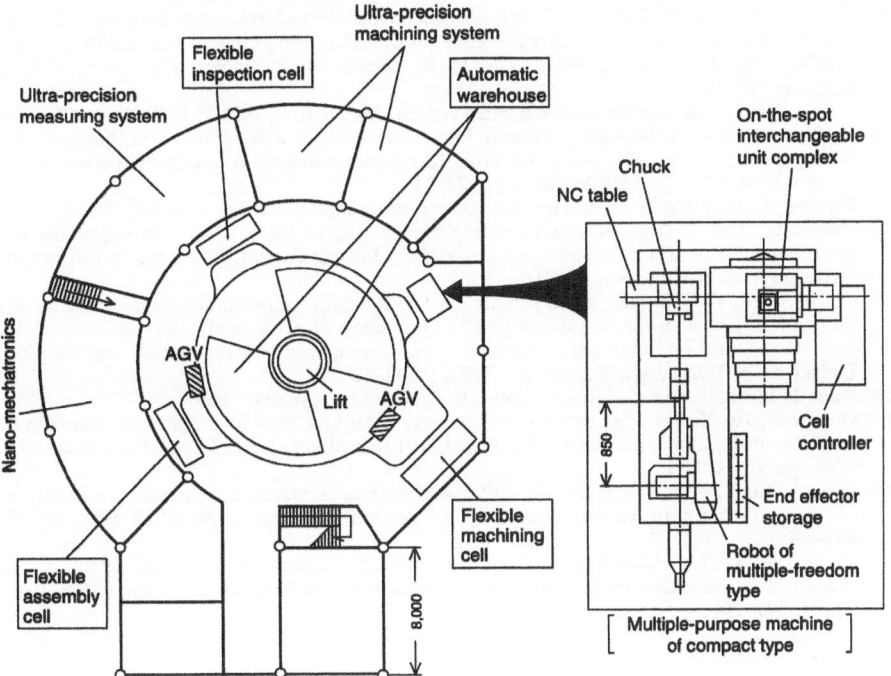

Fig. 1.18. Research facilities for thought model-based manufacturing (AIMAC) installed in Tokyo Institute of Technology.

a significant potential in the future production environment. At present, there are two representative variants, FCIMS of global network type and THOMAS.

This chapter has described the relationships between FCIMS, HIBM and THOMAS, showing case studies on systems in operation, prototype and trial stages. The chapter then focused on the basic research activities concerning FCIMS of global network type (culture of manufacturing), and THOMAS. The reader should thus have gained a perspective of human-intelligence-based manufacturing from this chapter that should prove a suitable grounding for the material that follows.

References

1. Weck M et al. Rechnerunterstützter Entwurf und Massnahmen zur Ausführung flexibler Fertigungssysteme. Industrie-Anzeiger 1974; 96 (74): 1683–1689
2. Ito Y. Amalgamation of human intelligence with highly automatised systems – an approach to manufacturing structure for the 21st century in Japan. In: Proc. 8th international IFIP WG5.3 conference (PROLAMAT '92), 1992 (to be published)
3. Ito Y. State of the art of advanced FMSs and FMCs worldwide. Trans of Mech Engng (IE Australia) 1991; ME16 (2): 91–99
4. Ito Y. The production environment of an SME in the year 2000. In: McGuigan K (ed) Flexible manufacturing for small to medium enterprises – a European conference. IFS Publications/Springer-Verlag, Bedford/Berlin, 1988, pp 207–234

5. Ito Y, Nishiyama M, Shima Y. Some of basic layout designs related to future production systems. In: Proc. international conference on manufacturing systems and environment – looking toward the 21st century. The Japan Society of Mechanical Engineers, Tokyo, 1990, pp 187–193
6. Ito Y. Conceptualizing the future factory system. Manuf Rev. ASME 1988; 1 (4): 252–258
7. Ito Y. A desirable production structure looking toward the 21st century – anthropocentric intelligence-based manufacturing. In: Proc. XI congresso brasileiro de engenharia mecânica, Promoção ABCM, São Paulo, 1991, pp 23–32
8. Brödner P. Skill based automated manufacturing. Pergamon Press, Oxford, 1986
9. Kolar MJ et al. Culture of manufacturing engineering in Japan – towards improving the international dimensions of engineering education. Project Report, University of Pittsburgh, Carnegie Mellon University, 1991
10. Moritz E, Ito Y. Computer aided production management – innovation and design/compatibility of technical problem solving patterns in Japan and Germany. In: McGeough JA (ed) Proc. 6th international conference on computer-aided production engineering, University of Edinburgh, Edinburgh, 1990, pp 11–18
11. Buur J. Mechatronics design in Japan. IK Publication, Lyngby, 1989
12. Chen MF, Ito Y. Investigation on the engineer's thinking flow in the process planning of machine tool manufacturer. In: 13th NAMRC proceedings, NAMRI of SME, Dearborn, 1985, pp 418–422
13. Ihara T, Ito Y. A new concept of CAPP based on flair of experienced engineers – analyses of decision-making processes of experienced process engineers. Ann CIRP 1991; 40 (1): 437–440
14. Ito Y, Shinno H, Nakanishi S. Designer's thinking pattern in the basic layout design procedure of machine tools – validity evaluation of thinking block. Ann CIRP 1989; 38 (1): 141–144

2 Design Principles of Work Organization and Skilled Labour in a Computer-Integrated Manufacturing Environment*

H. Hirsch-Kreinsen, C. Köhler, M. Moldaschl and
R. Schultz-Wild

2.1 Computerized Production Technologies and Work Design

2.1.1 Computer-Integrated Manufacturing and the Problem of Skills in Production

New types of rationalization strategies and a change in modernization policy have emerged in industry since the 1980s. The rapid pace of development in microelectronics and information technology promises that the technical resources to attain a new level of optimization between high productivity and high flexibility, which have been contradictory goals up until now, are now available. Due to the system character of these technologies, accompanying rationalization patterns are supposed to be both far-reaching and of a "systemic" nature [1].

The results of our survey [2, 3] indicate that the diffusion of the new technologies progresses much slower than was predicted by their advocates. This can be attributed to their high costs, lack of flexibility, high implementation risks, technical problems with interfaces between components in both hard- and software, and various other reasons. On the other hand, our observations suggest that in the early 1990s many companies are on

* This chapter is based on a summary of findings provided by a number of survey and case studies conducted at the Institute for Social Research (ISF Munich) originally written by the authors on behalf of the Institut Arbeit und Technik, Gelsenkirchen, FRG, as part of the FAST-project "Prospects of Anthropocentric Production Systems". These research projects deal with the scope for design with respect to work organization and skilled labour when CIM technologies were introduced into West German mechanical engineering industries. The authors want to express their special appreciation to Dipl. Ing. Eckehard Moritz for his numerous suggestions and helpful assistance in completing this chapter.

the verge of introducing new computer systems in the fields of production planning and control (PPC), some of them for the first time, and that there will also be an increase in the number of interfaces integrating various computer-aided design and manufacturing components (CAD/CAM). The trend definitely seems to be moving in the direction of more complex computer-integrated manufacturing (CIM) systems. All in all, a very dynamic development can be observed in the field of production-related computer applications.

Meanwhile it is widely agreed among experts that – at least in Europe – the concept of the highly automated "unmanned" factory will find only very limited applications within the near future and that this concept is at best suitable only for specific areas of highly standardized mass production. There is also a growing body of opinion stressing the limitations of the long-prevailing Tayloristic model of manufacturing organization and work structuring [4, 5]. This rationalization pattern is based on a distinct division of labour and job specialization on the shop-floor, and means centralized planning and decision making. Tayloristic rationalization was particularly important for large industries and mass production during the 1960s and early 1970s in order to overcome manpower shortages in an expanding economy. This type of rationalization strategy has not only dominated in mass production industries but has also been an important guideline for mechanical engineering industries, characterized by customized and small-batch manufacturing of complex products.

According to an alternative rationalization pattern, essential preconditions for competitive factory structures in the future will not only consist of computer utilization and increasing CIM, but also of skilled, qualified manpower on the shop-floor [6]. "New production concepts" [7], and new factory structures are being advocated which take advantage of the existing skills and qualifications and which seek to secure the potential for innovation and the ability to adapt, particularly in the case of the often smaller and medium-sized companies. In the long run, it will only be possible to utilize the potential for flexibility and productivity offered by the new manufacturing technology through skilled workers. Skilled and co-operative work within production islands on the shop-floor is often mentioned in this context as an effective way to reintegrate and reorganize tasks and functions.

It is often assumed that technological innovations will more or less automatically lead to an increase in the demand for skilled workers. This trend has not yet become apparent [8]. The risk still exists that skilled work will be undermined if tasks requiring skills are limited through automation and centralistic organization. This undermining of skilled production work could threaten the flexibility and quality of manufacturing, which has played a key role in the success of the German mechanical engineering industries. This does not have to be the case, however, if rationalization strategies and implementation processes of new technologies utilize opportunities for supporting the structures of *skill-based manufacturing* [9].

The term "skill-based manufacturing" is used in various, sometimes contradictory ways. In the context of our analysis we understand by skill-based manufacturing a certain type of *decentralized work organization*, characterized by the integration of highly skilled shop-floor workers into planning and decision-making processes in manufacturing. This type contrasts to Tayloristic or centralized forms of work organization which stress the importance of engineering functions and are based on distinct forms of division of labour.

It is certainly not *technology* alone which determines the impact computer-aided systems have on the organization of work and the structures of industrial labour. However, the way this technology is configured and implemented must be regarded as an important factor in this respect. An *automation-oriented* type of technology would be characterized by complex and inflexible man–machine interfaces and aims to replace or determine human decision making, whereas a *human-oriented* technology is designed to support its users in planning and decision-making functions.

The spectrum of *technical components* on the market for computer-aided manufacturing (CAM) or CIM has increased considerably. This holds true for design and process planning functions (CAD, CAP), for PPC, for manufacturing functions of machine control, tool management, workpiece handling and transport CAM as well as quality assurance (CAQ).

Basically, it can be assumed that there is a high degree of "elasticity" inherent in the various computer components as far as their basic design and configurations are concerned. Consequently, in theory, a corresponding scope for work organization and work design should also exist; in practical applications, however, this is not necessarily the case. As much as ever, there are many technical limitations – in particular, the often discussed interface problems – for combining virtually any number of different computer components. Besides that, user companies generally have more or less to accept the system concepts predetermined by the vendors and can at best modify them in a limited way in line with their interests, since they do not possess the necessary resources, i.e. funds, know-how, qualifications, and time. Therefore, due to specific technical features peculiar to the various system concepts, the scope for work organization and work design is likely to be substantially limited for the individual user.

Thus, CIM components, systems, and concepts can either create scope for innovating the organization of work or restrict the possible alternatives for a company. CIM components on the market can be classified into:

1. Those based on strongly centralized concepts which aim at a distinct division between conception and execution, i.e. between the functions of planning, control, and monitoring in planning departments, and work executed on the shop-floor.
2. Those which are more open and flexible in regard to different types of manufacturing organization and work structuring.

This chapter attempts to provide information on this dichotomy with regard to various PPC, CAD/CAM, and flexible manufacturing systems (FMSs).

2.1.2 Three Models of Work Organization

The implementation of computer-integrated components and systems generally has no massive direct impact on work organization. However, there are several associated factors, both inside the company and external ones, which suggest that there is a broad spectrum of possible developments in the area of manpower policies in the future.

The interaction of these and other factors does not compel company strategies in regard to work organization to move in any specific direction. On the contrary, there are many different ways industrial manpower strategies may develop. According to different forms and degrees of division of labour, three basic models may be identified.

Skilled and Co-operative Production Work. This concept (Fig. 2.1) implies a considerably low degree of division of labour – on the plant level between planning and engineering departments and the shop-floor as well as between jobs for shop-floor workers. This model presupposes the existence of a work force of qualified workers with skills of a relatively homogeneous level. Workers get assigned a whole package of various planning, monitoring, and operational tasks and functions ranging from process planning and programming to actual feeding and maintaining of the machine tools. The execution of these tasks is based on the principles of self-determination and co-operation. The concept of skilled and co-operative production work has been most widely implemented in the various forms of *group work*. Such work groups, e.g. system operators in an FMS, are able to work fairly autonomously. Sometimes these groups are integrated into the superordinate functional context of the manufacturing process according to the black box principle, i.e. as a semi-autonomous "plant within a plant".

While, especially in view of the increasing flexibility demanded by the diversification of market needs, there is good reason to assume that industrial work organized on the basis of skills is the only form which will be able to guarantee long-term economic efficiency in certain types of production (high quality, small series, complex products), it remains doubtful that this type of work will establish itself as the main trend, bearing in mind the current and foreseeable future conditions. Two possible obstacles are:

- Problems with the supply of skilled and qualified production workers for industrial jobs on the labour market.
- The lack of technical components and systems to accommodate such forms of work.

Other obstacles created by specific company structures should not be underestimated: e.g. the risks involved in abandoning well-established organizational structures and the associated shift from the established balance of power and interest constellations, and the extent of additional expenditure due to the necessary reorganization and further training.

Computer-Aided Neo-Taylorism. Unfavourable circumstances like those mentioned above may foster this alternative; its central feature is the further development and efficient enhancement of the hierarchical, functional and skill-related division of labour. Planning, programming, and scheduling are carried out exclusively by engineering departments. Service functions such as maintenance, repair, quality assurance and the management of equipment are systematically removed from the shop-floor and handled by specially trained skilled workers and technicians. The functions which remain on the

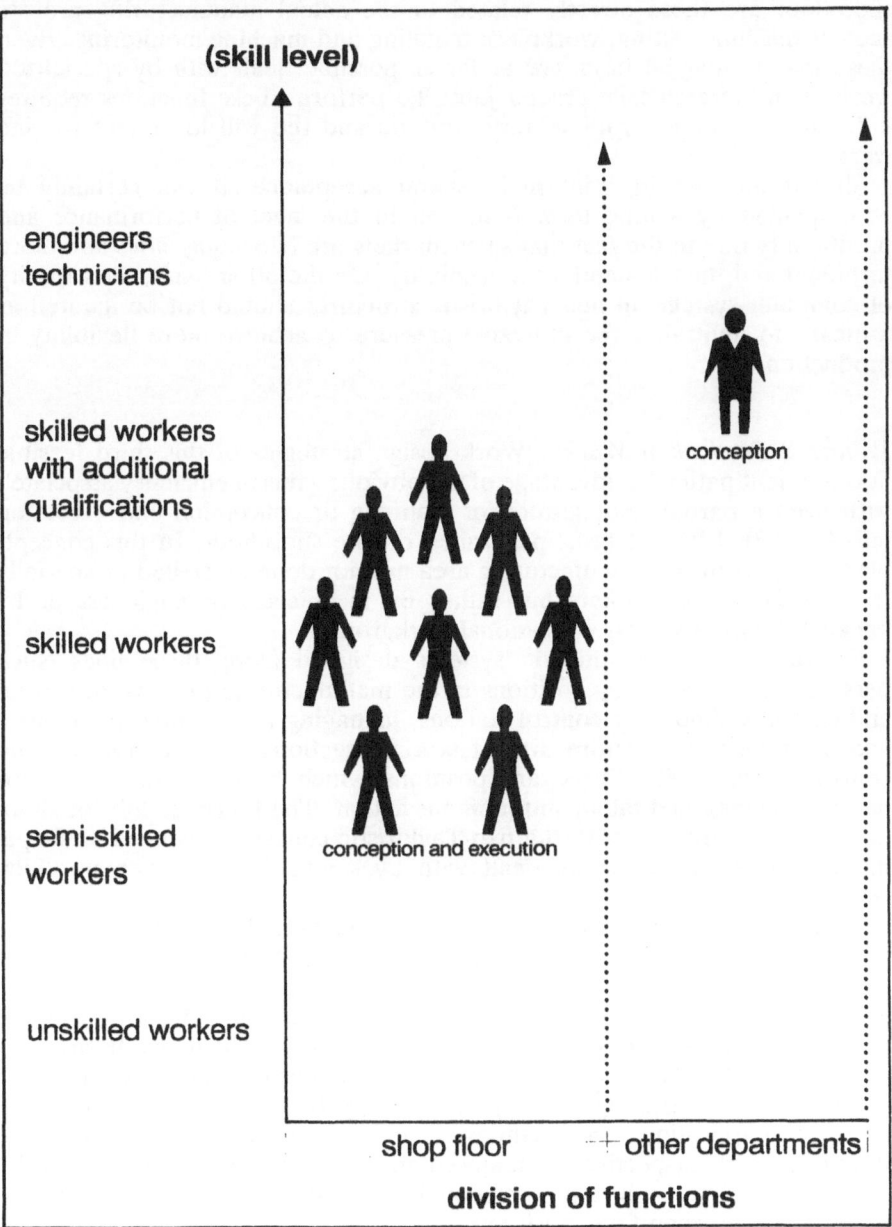

Fig. 2.1. Skilled and co-operative production work.

shop-floor are those directly related to the actual manufacturing process such as machine setting, workpiece handling and machine monitoring. Even these mainly manual tasks are as far as possible dealt with by specialized workers in hierarchically graded jobs. To perform these functions requires nothing more than a rudimentary aptitude and the will to endure routine work.

Should this development find general acceptance, it will certainly be accompanied by a long term reduction in the level of performance and profitability due to the fact that sales markets are becoming more and more turbulent and thus demand more flexibility. On the other hand, the capacity of computer-systems in neo-Tayloristic structures should not be ignored as a means to neutralize the objective pressure to achieve more flexibility in production.

Polarized Production Work. Work design strategies of this third feasible development path take advantage of the obvious gains in efficiency associated with even a partial reintegration of planning or conception and execution into the actual line of tasks performed on the shop-floor. In this concept, planning jobs in the manufacturing area are not done by skilled or specially trained shop-floor workers but rather by technicians or engineers or by personnel with a similar educational background.

On the basis of production systems designed along these lines, such personnel holds new key positions in the manufacturing process, at central and complex shop-floor control stations, managing for example the control and monitoring of entire manufacturing sections, or supervising and controlling an FMS. These are positions which have a large scope for decision making and taking autonomous action. The functions left for shop-floor workers are – as with the neo-Tayloristic concept – reduced to simple manual tasks which can be dealt with by semi-skilled or even unskilled workers.

This option would seem to have a considerable chance of finding acceptance particularly in comparison with skill-based and co-operative production work, especially in industries with no tradition of skilled workers. There are significant benefits with regard to flexibility and efficiency through a partial reorganization of the functional division of labour, while the risks of departing too far from the evolved and established company structures can be avoided. Although this option is a highly problematic one in terms of social implications and manpower policies, it offers highly attractive developmental perspectives for many companies. Moreover, many companies feel that they can acquire the personnel required for the comparatively low number of key shop-floor positions.

The following sections attempt to analyse how technology supports or impedes the development of work organization and structures of industrial labour in one or another of the directions mentioned above. We approach these questions by examining three lines of technology: systems of production planning and control (Sect. 2.2), the integration of CAD and CAM (Sect. 2.3), and FMS (Sect. 2.4).

2.2 Alternatives in PPC[1]

New means of PPC have been appearing on the market for some years now. One category of these are shop-floor scheduling systems or manufacturing control modules oriented towards the structures of comprehensive PPC systems. The other type are the "electronic control stations" (ECSs – *elektronischer Leitstand*) with a user-friendly graphic display based on the classic planning board (*Ganttchart*), interactive scheduling concepts and their own database on PC or workstation.

In this section, first the differences between the two types of manufacturing control systems will be explained. Then, the technical and organizational forms of employing the respective technology lines will be analysed to demonstrate the impact of different technology lines for alternatives in work organization.

2.2.1 Development Lines of Production Control

At the beginning of the 1990s, more than one-third of all the companies in the capital goods industry of West Germany were using computer systems for PPC. In the mechanical engineering and electrical engineering industries this figure was 40%. Similar to other types of computer technology, PPC systems are employed more frequently by large companies than by smaller ones. Correspondingly, 80% to over 90% of companies with 500 or more employees use PPC systems (see [2], p 35, and [3]).

Manufacturing control and in particular detailed shop-floor scheduling was a weak point of PPC systems for a long time. This applies in particular to the production of small and medium-size batches in mechanical engineering with its high degree of unpredictable and "unscheduled" disruptions, where short term rescheduling is often necessary, and where generally a high level of process complexity exists.

Improved performance and lower prices for hardware, and the further or new development of software components have led to an upsurge in innovations in PPC systems since the early 1980s. One result of this development is the differentiation of the system architecture into three levels:

- PPC (from sales through to job release)
- Manufacturing control – scheduling (from job release through to detailed machine tool utilization plans)
- Manufacturing control – execution (from real-time scheduling, job

---

[1] The following arguments are based on several sources: (1) discussions with experts in research institutes and with producers of manufacturing control systems; (2) some 25 case studies (conducted between 1987 and 1990) in companies of the metal industry with long experience in the use of PPC systems; (3) a survey conducted in the summer/autumn of 1989 with the 11 producers/vendors of ECS known at the time; (4) an expert report of the IWI Saarbrücken on the subject of the technical features of these systems on the market and their suitability for semi-autonomous work groups and production islands [10]. A detailed account of the research findings can be found in [2] and [11].

sequencing and process control through to production data acquisition (PDA) and monitoring)

Spectacular changes can be observed on all three levels. In particular two lines of development can be identified. These are ECSs on the one hand, and central shop-floor scheduling systems or modules on the other.

ECSs are designed for the needs of small and medium-size batch production. The functional range and databases are built up successively. Automatic scheduling using different heuristics can also be provided, but the emphasis remains on the dialogue. The user interface is easy to handle and is readily accessible by "part-time schedulers" such as foremen, group leaders, and production workers. ECSs are technically and economically compatible with various organizational forms of production control. Due to their hardware and software features, they are suitable both for centralized forms of production control (by specialized personnel scheduling large units) and for decentralized forms (by foremen, group leaders, or production workers in small units). Consequently, they may be described as an *open* technology line.

Central shop-floor scheduling systems, mostly sold as modules of comprehensive PPC systems, differ from ECSs principally. Due to their technical features, this line of automation-oriented technology is especially suitable for organizational forms of *centralized manufacturing control*. Equipping a large area with small units of interactive terminals would normally be more than the capacity of the hardware could handle, and would result in response times unacceptable for real-time control. Parallel scheduling of many small units is frequently not provided for in the software structure and would require large-scale, time-consuming, and expensive adaptations. The cost of the hardware and software is an argument against seeking to solve such problems by purchasing a larger number of these systems. In addition to restrictions imposed by the hardware and software, the *user interface* is a considerable obstacle to a utilization in decentralized set-ups. Because of the time needed to master the functional complexity and the user interface, it would seem more logical to have full-time system operators; however, scheduling the work for a foreman's unit or a work group does not make up for a full-time job. On- and off-the-job training take time; acquired knowledge is quickly forgotten if the operator works only sporadically with the system. The gap between the knowledge and experience of first line management and production workers and the abstract operational logic of the systems is too big.

In line with the distribution of data and functions on the various PPC levels, three technical concepts of production control can be differentiated today (Fig. 2.2):

- In the first concept, a centralized system for manufacturing control dominates. The centrally calculated schedules extend to the job sequences of individual machines or workplaces.

- In the second concept, ECSs are employed to supplement a central system for shop-floor control or module. Since the modern central systems have a high scheduling frequency (mostly daily) and a high accuracy, the remaining scope for decisions at the control station is relatively small

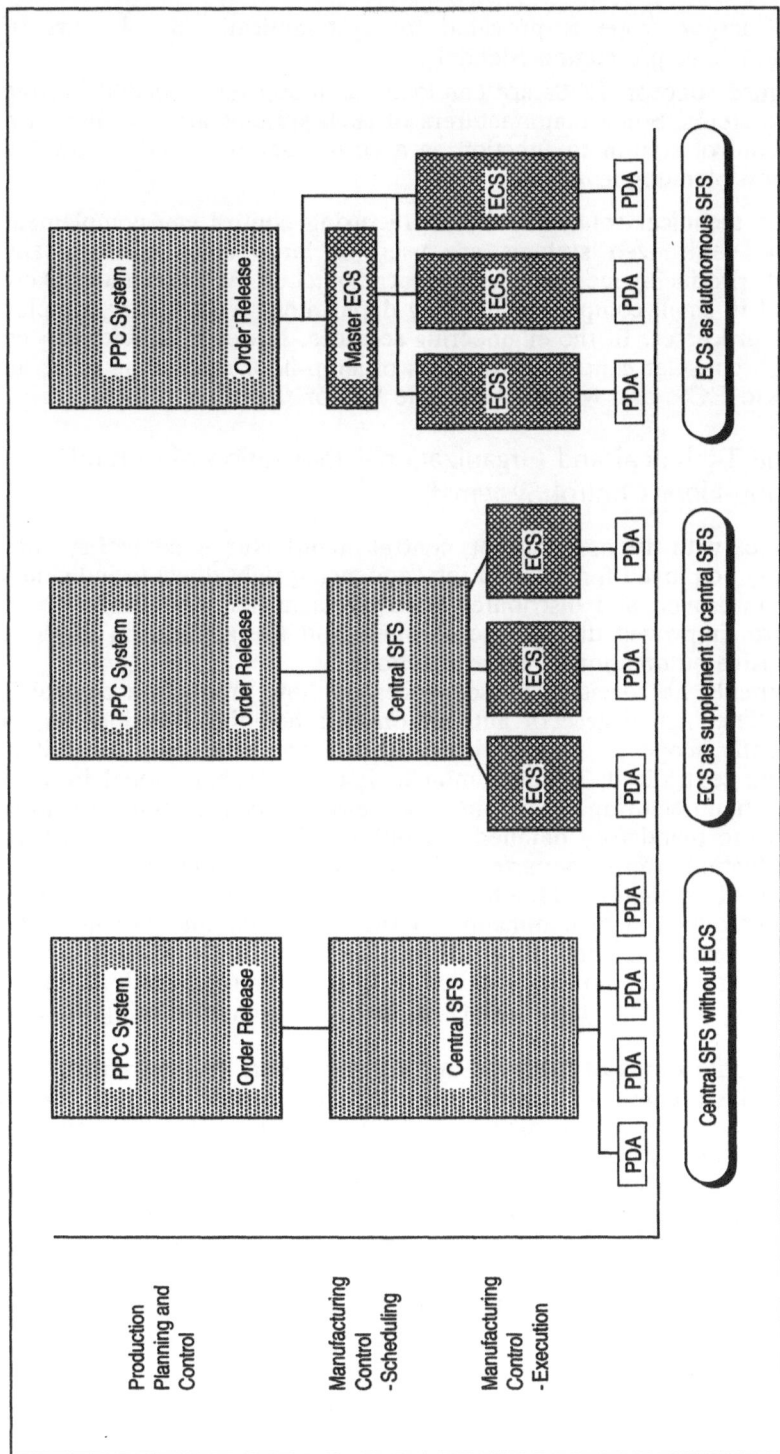

Fig. 2.2. Technical concepts of shop-floor scheduling systems (SFS) and the application of electronic control stations (ECS).

unless a larger scope is provided for systematically in sub-areas of production (e.g. production islands).

- In the third concept, ECSs are employed as independent manufacturing control systems. Some manufacturers of such systems are working on a master control station to function as a co-ordinating system below the production planning level [10].

These three technical concepts of manufacturing control can complement each other: centralized systems are used in large companies/company departments producing medium-sized or large batches; ECSs are more likely to be found in small companies/company departments with highly complex production processes. In the engineering sciences, however, lobbies can be found which consider centralized systems of shop-floor scheduling to be an alternative to ECSs and which favour one line or the other [12,13].

2.2.2 The Technical and Organizational Integration of Central Shop-Floor Control Systems[2]

In accordance with the first concept, central manufacturing control systems are generally employed for detailed job sequencing right down to individual machines. Improved and distributed hardware, further modularization of the software, improved dialogue techniques, and the application of PDA allow extensive automation of these functions.

This normally has decisive consequences on how production control is organized. The high degree of automation of scheduling functions leaves relatively little scope for human intervention. The remaining scheduling tasks can be centralized. It is common to find the organizational form of centralized manufacturing control in which either a central control station determines the mandatory detailed scheduling for large production areas, or the production control department is responsible for the entire factory. On the basis of detailed machine tool utilization plans, this central authority issues mandatory control commands to the shop-floor and monitors the progress of work on the job. First line management on the shop-floor and production workers are no longer called on to perform organizational tasks; therefore they will lose their ability to keep the manufacturing process flexible.

Organizational forms of this type are *not necessarily* the consequence of the technical structures of central manufacturing control systems. Some latitude in decision making may be conceded to foremen and machine operators by laying down less detailed schedules.

However, a decision once made for employing a central shop-floor scheduling system seems to presuppose centralized forms of work organization as well. This is without a doubt the general mainstream trend in applying modern PPC systems. There are many economic, social, and political factors supporting these types of Tayloristic concepts in production control [8]. It

---

[2] With regard to this and also as background to the following comments, see the comprehensive sociological analyses of Manske on production control in mechanical engineering [14,15]. See [16] in particular, and [2], p 171ff for analyses on organizational forms of production control.

should be emphasized here – in the context of technical constraints – that due to their hardware and software limitations, centralized systems are not suitable for supporting decentralized decision making. With the increasing economic importance of short throughput times and high reliability of delivery dates, it makes little sense for many companies – especially once they reach a certain size – to curtail the scheduling accuracy of these systems consciously and do without computerized tools supporting their manufacturing control.

Centralized systems for manufacturing control which determine job sequences down to the individual machine tool are hardly reconcilable with forms of *skilled and co-operative production work* with decentralized scope for decision making. They are suitable in particular for forms of computer-aided neo-Taylorism or polarized production work.

Only the supplementation or substitution of central shop-floor scheduling systems by ECSs opens up a degree of latitude for organizational design. For this reason, the following remarks focus on the application of this relatively new, and still little-utilized technology.

2.2.3 Technical and Organizational Integration of ECSs

A survey[3] of ECS producers in West Germany shows that the main area of application is typically in the medium and large-scale mechanical engineering companies with customized and small-batch production organized in job-shops. About two-thirds of the surveyed control stations fall into these categories (Fig. 2.3). In 14% of all cases, ECSs were employed in production islands. Almost all ECSs are linked to PPC systems [11].

Control stations have proven their worth as effective and up-to-date information, scheduling, and control systems. Obviously they achieve a degree of accuracy and precision far greater than that of the older PPC systems with their weekly batch runs, but also of the modern systems with central scheduling modules and daily batch runs.

Almost two-thirds of the ECSs installed today are used according to organizational structures of *centralized* manufacturing control (Fig. 2.4): They are operated by specialized personnel responsible for several foremen's units and job-shops. Since the scheduling accuracy of ECSs is generally very high, the foreman loses many of his control duties in such cases.

All together, only slightly more than one-third of all the ECSs surveyed operate in organizational settings of *decentralized* manufacturing control:

- Twenty-eight per cent of all ECSs are used as decision-making support for the foreman; the number of controlled work places is less than 50 in half of these cases, and between 50 and 100 in the other half.

- Only 8% of all ECSs control smaller units. The system is then used in

---

[3] In the summer of 1989, according to the 11 surveyed producers and vendors of ECSs in the Federal Republic of Germany, there were some 190 systems in use. A further 130 installations were planned to be in operation by the end of 1989. For these, manuals already provided in concrete form the contours of how they were to be used. This total of some 320 control stations is the basis of the following analysis.

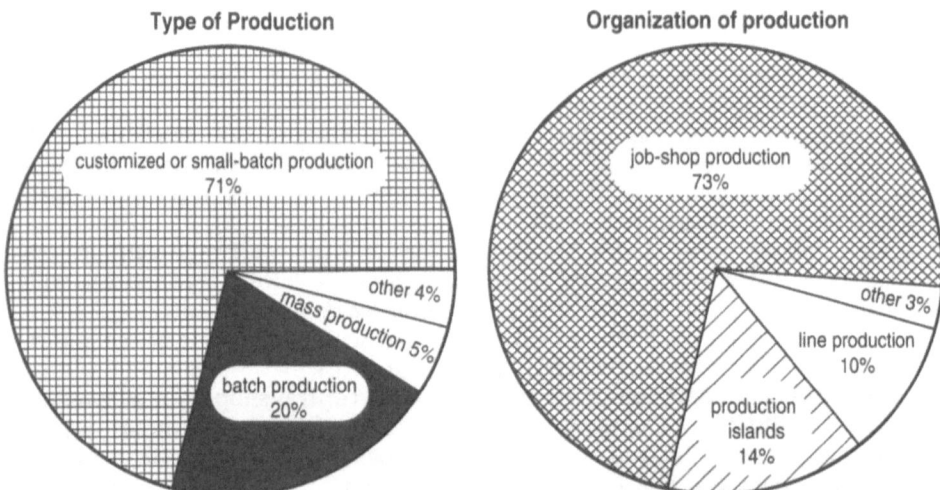

Fig. 2.3. Electronic control stations – areas of application (320 cases).

the context of group work (e.g. in a production island for the purpose of creating detailed machine tool utilization plans) and is normally operated by the group leader.

The organizational forms of *centralized* manufacturing control are still prevalent in the utilization of ECSs. The producers predict a trend towards decentralization, however. Among the 11 companies surveyed, eight, including the leading two producers in terms of sales, expect that forms of decentralized manufacturing control (by a foreman or a semi-autonomous group) will prevail when employing ECSs in the future. Two producers advocate both centralized and decentralized development paths. Only one producer views the centrally employed control station as the ideal solution.

2.2.4 The Participation of Production Workers in Manufacturing Control

The organizational integration of ECSs does not remain without repercussions for the division of labour between production workers, foremen, and specialized control personnel regarding detailed scheduling. Production workers have always been and still are integrated in shop-floor scheduling in many areas of mechanical engineering as they are involved in the selection of the "job bundle" assigned to them and decide on the sequence within the bundle [14,15]. As a result, decentralized logics become important (the interests of wage stabilization in piece-work systems, optimization of set-up times, stress reduction, job variation, etc., may be mentioned here); they can sometimes counteract centralized logics, but sometimes support them as well.

This decision-making latitude of production workers will undoubtedly be curtailed by the modern manufacturing control systems which operate with increasingly accurate and detailed schedules. The application of ECSs reinforces this trend by simplifying the job sequencing for individual

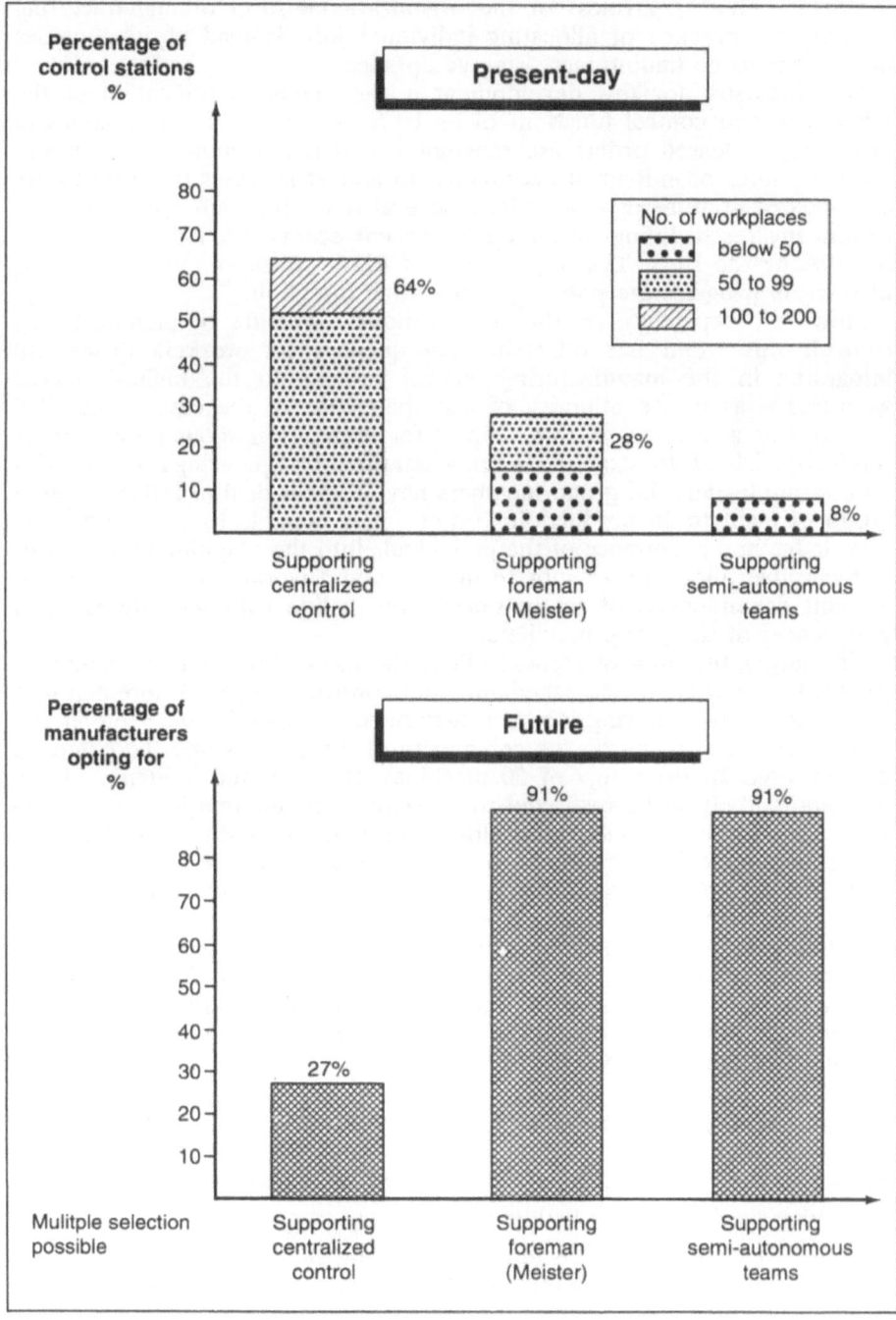

Fig. 2.4. The organizational integration of ECS.

machines. Thus, regardless of the organizational form of manufacturing control, the practice of allocating individual jobs instead of job bundles now seems to be finding increasing acceptance.

One indicator for this development is the extensive utilization of the scheduling and control functions of an ECS. At almost all the systems in use today, released orders are scheduled in detail including starting and finishing times of individual operations. In almost all cases these schedules are updated continuously or at least several times per shift and day. As a result, any rescheduling, prompted by urgent orders, breakdowns, etc., is handled by the ECS. The high degree of PDA utilization and monitoring of work in progress are pointing in the same direction.

However, depending on the organizational structure of manufacturing control, this trend has different consequences for workers. They are integrated in the manufacturing control process to the highest degree wherever – as in the minority of less than 10% of the cases – an ECS functions as a decision-making support for semi-autonomous groups, as in production islands for example. Even when the job sequencing is determined by a group leader, the group members have a great deal of influence. In a group of eight to 15 production workers, it is possible to incorporate the knowledge and experience of the individuals into the planning by means of a favourable allocation of jobs to machines and workers (e.g. taking into account dependencies of set-up times) and in line with the interests and preferences of the group members.

The larger the area of responsibility, the more difficult it is to involve production workers in the scheduling and control process. A foreman in a mechanical manufacturing job-shop may have to supervise between 30 and 100 workers depending on the complexity of the process and the company organization. In the range of 50 machines or more, the influence of the workers is likely to be restricted to subsequent minor revisions of the job sequences. Only in smaller units closer forms of co-operation are feasible.

Production workers are likely to have the least influence when ECSs are used within organizational structures of centralized manufacturing control. This is the situation in just about two-thirds of the systems where the scheduling of up to 200 machine tools is carried out. Even partial integration of production workers in the scheduling process would overtax the information processing capacity of the control operators, and for that reason will usually remain an exception (unless workers have been given some latitude through the allocation of "job bundles").

2.2.5 Development Trends

To summarize, two concepts of the technical and organizational integration of control stations can be identified today:

- In the first concept, ECSs function as supplements to central shop-floor scheduling systems. They tend to be but stopgaps for centralized planning and scheduling.

- In the other concept, the ECSs are operated technically with a large degree of scheduling latitude as independent systems below the production planning level; organizationally, they support decentralized units.

In the first concept, ECSs are simply a means of transferring and adapting central scheduling to the shop-floor; little scope for decision-making is left. This type of utilization will possibly become obsolete in the further development of centralized manufacturing control.

In the alternative concept [17,18], ECSs are given a high degree of autonomy in scheduling functions (with planning terms of up to five days or even several weeks). Sometimes this autonomy is extended to process engineering and routing functions. The co-ordination between decentralized units (e.g. production islands) and the production planning level is taken care of by either central shop-floor control systems or a master ECS. From an organizational viewpoint, the control station is allocated either to a foreman or to a semi-autonomous group. The central monitoring of work in progress is characterized by the motto "as much as necessary and as little as possible".

Both concepts and development trends signify ranges of the application of ECSs which vary depending on the size of the company, the type of production, the organization of manufacturing, and the company philosophy. A glance at the modern centralistic concepts can easily convey the impression that the mistakes of earlier PPC developments are being repeated. The scheduling and control potential is overestimated, valuable and irreplaceable skills of workers and supervisors on the shop-floor remain unutilized or are even lost. While the modern systems have pushed forward the boundaries of scheduling by yet another milestone, they have by no means solved the fundamental problem of complexity in manufacturing control in the area of batch production.

The application of modern PPC systems and strategies of production segmentation was able to bring about a considerable reduction (sometimes up to 50%) in throughput time in many companies. In mechanical engineering with job-shop production, however, the main application area of ECS, these strategies have specific limitations: below a certain buffer size, capacity utilization declines. If it is not possible to go back to alternative jobs when problems with the scheduled jobs arise (missing jigs, tools, NC programs, etc.), the machines will stand still. As much as ever, long idle periods for semi-finished products have to be tolerated if the more and more capital-intensive facilities are to achieve high operating times. If the average throughput times of parts requiring ten operations in mechanical manufacturing was three months in many enterprises prior to their modernization (e.g. production segmentation and new PPC systems), then six to eight weeks is considered a good result. With the exception of urgent orders, it is hardly possible to improve on a few days of idle time per operation when small parts are manufactured.

If just-in-time remains an illusion in batch production, then there will still be much scope for optimizing the allocation of jobs, manpower, and machines by the foremen and by production workers to reduce set-up times, to improve quality and throughput speed, to meet deadlines, and to increase work satisfaction and motivation. Much of the information necessary for decentralized optimization processes cannot be easily computerized. This also applies for production workers' control knowledge based on experience and sometimes mental pictures which cannot be communicated easily. What cannot be computerized at all are unforeseen occurrences in the manufacturing itself, which may cause changes in the production process, or require human reasoning capacity. This would include, for example, changes in the "daily form" of workers and fluctuations in temperature to the extent that they influence the operation of individual machines and workpieces. Changes in the production context which lead to a change of planning objectives on the shop-floor are very important. When conflicts regarding deadlines occur, the relevance of customer orders has to be weighed; different strategies are required in instances of operation above capacity and in cases of light loading.

The quality of and speed in dealing with an order depends very much on the degree to which the interests of the production workers have been taken into account. This takes in far more than wage stability and regulation of stress; it means preventing monotony by diversifying the work and ensuring interesting and challenging tasks. The knowledge and experience and the direct personal relations between first line management (foremen, group leaders) and production workers remain factors of great significance for shop-floor scheduling, particularly in the context of flexible production processes; they can therefore not be considered as marginal factors which will be made superfluous by new information technology [10].

ECS are an excellent means of utilizing and mobilizing this knowledge and experience.[4] The high degree of user-friendliness of the systems on the market predestines them for decentralized use. For this reason they should not be employed as a mere appendix to PPC and central manufacturing control systems but rather as active and important elements with their own scheduling areas in the overall system of PPC.

2.3 CAD/CAM Integration and Alternatives in Work Organization

The term CAD/CAM and the underlying integration technology have been around for more than 20 years. They date back to technological concepts which originated in the first half of the 1960s in the wake of numerical control (NC) development in the USA. Admittedly, these systems have spread only haltingly so far [2].

The integration path of CAD/CAM is very broad; it covers all technical aspects of production in a company: from engineering and design to process planning and manufacturing up to and including quality assurance. The breadth of its integration comprises the entire workpiece spectrum of a company and the associated machining processes. Consequently, CAD/CAM systems integrate the individual technologies of CAD, CAP, CAM, and CAQ. Together with computer-aided integration of the economically oriented functions in the handling of orders by PPC systems, CAD/CAM systems virtually form the backbone of a CIM system.

---

[4] For further information and analysis on the categories of knowledge and experience, see [19].

2.3.1 Basic System Alternatives

The following will examine the prospects associated with integrated CAD/CAM systems with respect to maintaining or extending skill-based manufacturing. As indicated above, various concepts exist here; these concepts offer in principle a broad scope for design alternatives especially with regard to work organization on the shop-floor. On the other hand, due to specific technical features peculiar to the various system concepts, the scope for work organization and work design is likely to be substantially limited for the individual user introducing the systems.

This holds in particular for CAD/CAM systems as they tend to link and integrate practically all the "vertical" process planning and technical design functions of a company by means of information technology. Consequently they affect the relationship between division of labour and co-operation at the same time. This concerns in particular the possibilities for organizing the internal division of labour in functional and hierarchical terms. Technical specifications influence the distribution of conception and execution. As a result, there is a limited degree of manoeuvrability both in the practical configuration of machinery and jobs on the shop-floor and in the related structure of skill requirements, which definitely affects the distribution of functions of technological process planning and of the programming of NC machine tools. These functions can either be executed in an organizational structure based on division of labour in one of the offices separate from the shop-floor (process planning or even design department) or else on the shop-floor itself where it can represent a focal element of skilled production work.

These correlations apply less for comprehensively integrated CAD/CAM systems, which the users could adapt to the specific requirements of their company. They do hold, however, for the relatively widespread "sub-lines" in CAD/CAM integration, CAD/NC and direct numerical control (DNC). Whereas in the various forms of the CAD/NC integration CAD systems are linked with NC programming systems for the common use of geometric data of parts, the DNC sub-line generally integrates NC programming systems with the various types of CNC machine control units via the use of a DNC computer. The core function of DNC systems is the on-line transfer of NC programs between the various system components.

For both types of CAD/CAM integration mentioned above, complete solutions are offered on the technology market. On the one hand, they always require adaptation, modification, etc. to the needs of the company. On the other hand, they offer specific technical features which are associated with a different potential with regard to work organization and job design. Differences of this kind can be found both between and within the individual sub-lines. Typical features from a viewpoint of information technology are e.g.:

- The concrete functional design of computer components
- The structure and interface of the user programs
- The form of data organization and data transfer via interfaces
- The respective hardware configuration

If one now attempts to analyse the current and foreseeable offers of

CAD/NC and DNC systems on the technology market, in a first rudimentary distinction two types of system concepts can be identified:

- A concept based on division of labour
- A concept which is flexible with regard to work organization and consequently open for shop-floor workers

These concepts suggest different ways of dividing the work functions of planning, adapting, and implementing processes between the processing planning departments and the shop-floor (Fig. 2.5). The various system concepts are both an indication of the underlying socio-economic conditions of the technological development and the structures of the technology market.

2.3.2 The Predominance of System Concepts Based on Division of Labour

The majority of CAD/CAM concepts end up maintaining the traditional division of labour between conception and execution within an organization of work differentiated according to hierarchic-functional criteria. This main path of CAD/CAM integration is obviously determined mainly by large computer manufacturers and system vendors.

There is a marked orientation to large-scale company structures with a high degree of division of labour and complex requirements for manufacturing; this is an indication of the particular impact of the American aerospace industry on CAD/CAM development (cf. [20]). This applies in particular for a number of specialized American CAD/CAM vendors who have had a strong influence on the Western European market for these systems, and partly dominate it. Some French developments in this field point in the same direction.

De facto, these system concepts concentrate on co-ordinating functions in design, process planning, and on the shop-floor more systematically than was possible with conventional methods. These areas are organized on the basis of a distinct division of labour and are traditionally separated from each other. In addition to co-ordinating functions, these systems establish a steady and efficient flow of information between the various departments. The central feature of these system concepts based on division of labour is that they integrate an NC programming system which supports office-oriented operation in some form or another but is not suitable for shop-floor operation.

As far as it is possible to assess the products available on the market, it would seem that at the present time all CAD/NC systems are heading in the direction of an enforced division of labour between planning and programming on the one hand, and execution on the shop-floor on the other. The possibilities of integrating standard CAD systems generally cover NC systems only, which are linked in one form or another to office-bound operation.

This applies to systems where an NC module is integrated into the software of the CAD system; this requires a relatively expensive and complex CAD station for programming, which is only suitable for office operation. Such a CAD station is too expensive for merely occasional use

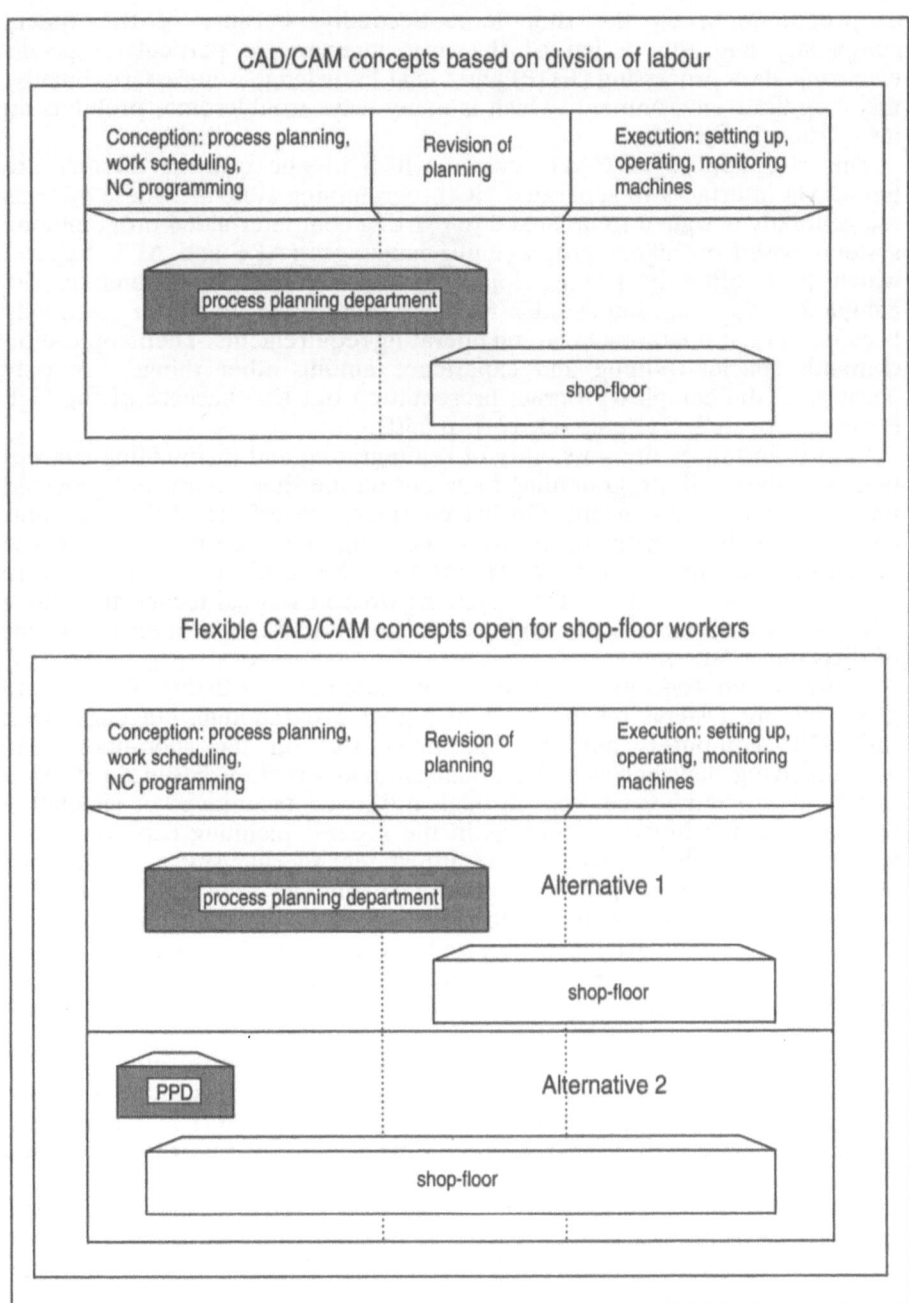

Fig. 2.5. Organizational alternatives of CAD/CAM integration.

in programming on the shop-floor. Secondly, because of the system complexity and the design of the user interface in particular, specific electronic data processing (EDP) and CAD knowledge is necessary. Finally, the shop-floor environment, which is many ways troublesome, prohibits an installation in this area.

This also applies to CAD systems which in one form or another are linked via interfaces to separated NC programming systems. These systems are generally designed to be linked to existing computer-aided programming systems based on higher programming languages (APT and APT dialects) which have often been used for a considerable length of time in the companies; these are not suitable for operation on the shop-floor essentially because of their operating logic and operating requirements. Their application demands special training and experience among other things, not only because of the complex program presentation but also because of the high error rate of these systems (cf. [21], p 55ff).

In both instances the possibility of reintegrating and maintaining conceptual, and above all, programming functions on the shop-floor is not provided for in the technical concept. On the contrary, the effects of these systems are more likely to mean transferring the programming functions onto the design department. From the viewpoint of work organization, scarcely more flexible is a whole series of DNC systems whose essential technical features also de facto support the preserving of existing structures with distinct forms of division of labour.

Many system concepts seem to rely on integrating existing office-bound programming systems on the basis of higher programming languages with the DNC computers and NC machine tools on the shop-floor. The accompanying deficiencies of an organizational structure based on division of labour are minimized, e.g. through improved techniques of simulating programmed machining operations in the process planning department, as well as by a number of control and adjustment mechanisms to co-ordinate between the (source-) programs provided from the offices and optimized on the shop-floor. Associated with this is the frequent possibility to access stored NC programs which may be tailored to meet the company's needs. While, e.g. the NC programming office has access to all the programs in a central storage, shop-floor workers might have only restricted access to programs "released" by the NC office.

Beyond this, DNC also allows for the utilization of more or less comprehensive functions of PDA and machine data acquisition (MDA). This is often considered a prerequisite for profitable DNC operation. However, it only makes sense within the framework of a work organization based on a distinct division of labour. The goal is to increase the control and planning ability of the process planning and engineering departments "above" the shop-floor. Furthermore, a number of advanced systems still in development envisage the integration of an ECS, centralized with regard to the work processes on the shop-floor, in which additional scheduling functions such as process planning or manufacturing control can be carried out as well as programming and its optimization.

Finally, DNC systems are also conceived in various ways as more or less automatic feedback control systems monitored by a shop-floor ECS or the process planning department. Data transfer is regulated according to the respective stage of the work in process, which is passed on via monitoring systems to a DNC or ECS. Convenient numeric control systems at machine tools which can in principle be used for shop-floor programming also assume in this configuration the central function of automated data acquisition (PDA) and data exchange.

2.3.3 Growing Significance of CAD/CAM Concepts Suited to the Shop-Floor

The application of CAD/CAM systems based on a division-of-labour concept is encumbered with problems for a variety of reasons. There are two main barriers: first, in a large number of smaller and medium-sized companies, which have only a limited division of labour, such system concepts can only be introduced with difficulty and with a certain element of risk. This applies, e.g. to the wide spectrum of tool and die manufacturing. Second, the intensified dynamics of the sales markets call for flexible company structures, and larger companies in particular still often find it difficult to reconcile their prevailing patterns of division of labour and hierarchy with such structures.

Due to these and other reasons, in recent times more and more alternative CAD/CAM systems have been developed which can be used flexibly with regard to work organization and work design. They offer companies considerable scope for shaping their work organization; in particular they allow for the execution of planning and programming functions on the shop-floor.

These systems which so far can be called sidelines of the CAD/CAM development path are being promoted by smaller, specialized software and system vendors, partly in co-operation with mechanical engineering companies. These concepts are based on specialized production-related know-how which is incorporated into the development of adaptable systems as well as on the accelerated development of hardware which greatly facilitates increased flexibility in system concepts and their departure from earlier "mainframe solutions".

There is no doubt that particularly complex machining conditions, e.g. the production of complex surfaces by a multi-axle process which is difficult to determine geometrically, still demand a specifically office-oriented design of CAD/CAM systems. These systems need for example the components already listed, like complex programming systems, elaborate graphic terminals and possibly the connection of the system to a mainframe computer. However, these considerations hardly apply any more for most less complex machining processes. The development is shaped by the increasing application of personal computers, and of efficient microcomputers, the so-called workstations.

There are various indications of a diversification of *CAD/NC integration*, which so far has been predominantly office-oriented. For some time now systems "tailored to the shop-floor" have been on the market. They consist of completely integrated hardware and software for a computer-aided workplace; they incorporate CAD and programming functions and can be linked with CNC machine tools to an integrated CAD/CAM system. Due to relatively simple functional scope and, more particularly, the convenient user interface, the application of these CAD/NC systems is obviously also possible next to the machinery on the shop-floor and close to the manufacturing process. In this case, shop-floor workers can sketch the contours of parts and write the corresponding NC programs themselves.

Furthermore, there are concepts which aim to *increase the flexibility* of previously centralized systems, but these are only in the initial stages of development. First and foremost, considerations directed towards traditional CAD/NC systems on an APT basis should be mentioned; these rely on the shop-floor-oriented application of computer systems often already utilized in the company, and the corresponding data files. The central prerequisite for this concept is considered to be the "opening" of interfaces between system components; the various existing forms of integrating CAD and NC systems via interfaces open up a range of new opportunities. Generally the goal is to increase the possibilities both for combining programs, data, and the system components, and for facilitating communication between them.

Secondly, more recent concepts of a *direct integration between CAD and the shop-floor* should also be mentioned for which a dedicated programming system is not required. The design data are intended for direct utilization in process planning. While in the PC version, the office-oriented organization of planning and programming is not impossible in principle, direct integration of CAD and CNC machine tools on the basis of a drawing "appropriate for a CNC-application" [22] allows these functions to be executed completely on the shop-floor. Such integration concepts were introduced for turning operations.

In the case of *DNC systems*, technical concepts which may be described as flexible or shop-floor-oriented have been available for quite a long time and they are constantly being improved. These concepts comprise an NC system which is not designed exclusively for office operation but is rather a programming system in which the operating logic and the necessary hardware allow, with respect to work organization, a highly flexible configuration of the system via terminals, i.e. workstations and integrated CNC controls. These may be relatively simple so-called program editors which have a limited range of application, or elaborate and universally applicable interactive graphic systems. These DNC systems can be extended step-by-step without obvious difficulties, and adapted to specific company needs, which means that they are compatible with any kind of organizational environment.

A central prerequisite for the flexibility of such system concepts in terms of work organization is that the various system components have a *uniform user interface* either directed towards the specific control logic of CNC controls or as a universal interactive graphic system, enabling the simulation of programmed machining operations as well. In addition to the programming functions, these systems allow for various additional planning functions such as machine set-up planning which reinforces the applicability under different work environments, in particular the extensive use of the system by shop-floor workers.

2.3.4　Harmonization of System Concepts?

Without a doubt, the CAD/CAM development in the future will be shaped, at least partially, to an increasing degree by system concepts which are flexible with regard to work organization. In particular the further development of programming systems and correspondingly the reduction of differences between office-bound and shop-floor-oriented programming systems will probably be of significance here. They will be characterized by a progressive homogenization of operator modes following the further development of interactive graphic techniques.[5] With this presumably lengthy process of CAD/CAM development, the current trend towards modern and extremely flexible CNC controls which run on convenient, directly programmable small computers will be duplicated.

If considerable limitations on the application of shop-floor-oriented systems can be overcome by (re-)integrating planning functions on the shop-floor level, more and more restrictions may be imposed on shop-floor operation under certain circumstances through the progressive establishment of networks to CAD systems. As has been shown, practically all the currently available computer integration paths encompassing CAD systems are only suitable for application in an office-like environment. Shop-floor-oriented integration with CAD systems is still in the initial stages of development and it remains to be seen to what extent this concept will be applicable in the future.

Decisive for the development prospects of these and other CAD/CAM concepts that are flexible with regard to work organization are, however, not so much the basic problems of information technology which are still to be overcome, but rather the future strategies and interests of the system manufacturers and other institutions involved in developing these systems. In this respect, the development prospects of computer-integrated production systems can presumably be described as relatively undetermined. At present, an interaction between various development conditions can be observed.

On the one hand, it can be assumed that system development is becoming increasingly internationalized and standardized under the aegis of the major internationally active, mainly American computer manufacturers and software vendors. Office-oriented system concepts and those encompassing CAD integration elements will be similarly influenced by this trend. An increasing application of flexible systems will presumably only occur within the framework of general EDP development in the direction of increasing user-friendliness, etc.

On the other hand, further differentiation of the flexible systems available to solve a whole range of problems of user companies can be envisaged. This development will probably continue to be borne by specialized EDP vendors and mechanical engineering companies. These system concepts will not only fit the needs of companies operating in specialized market niches,

---

[5] In Germany, the government-sponsored development project "WOP" ("Shop-floor-oriented programming procedures") was trailblazing in this respect. It has provided and continues to provide an impetus towards homogenizing and, to a certain extent, standardizing universal programming procedures.

such as tool and die manufacturing, but will also be suitable for the structures and requirements of the majority of medium-sized and smaller metal-working companies. Of course, the resources of both this developer group and the associated user companies are limited, so it is questionable whether they will be able to cope with the rising costs of system development in the long run. Experts assume that in the future there will be a closer co-operation between manufacturer groups, the large computer manufacturers, and system vendors. It is difficult to predict at the present time what effect this will have on the further development of system concepts flexible for varying organizational environments.

2.4 Flexible Manufacturing Systems – Design Possibilities Re-examined

FMSs cover a whole spectrum of technologies: machining, material flow, scheduling, and control technologies. They comprise elements of both integration paths: the "horizontal" planning and control of orders and production flow (via PPC systems etc.) as well as the "vertical" planning and control of production from design and programming, to machining and quality assurance (via CAD/CAM systems), which are examined more closely in Sects 2.2 and 2.3, respectively. In an increasing number of cases, FMSs are also linked externally to superordinate control systems. An FMS can be described as a highly advanced "small-scale CIM" – production and organization technology in one system. Both control technologies are important for determining the scope and limitations of factory and work organization.

2.4.1 FMS as Office-Oriented Automation

The core element and technological basis of any FMS is the CNC machine tool. In the 1960s and 1970s, triggered by the advances in the numerical control technology, manufacturers producing customized and small-batch goods managed to narrow the productivity gap to large-scale production and to increase the degree of automation in machining processes. Efforts were directed towards automating the periphery of machine tools (rotating tables, turret heads, tool and pallet changers). With the introduction of the FMS, the main objective followed by nine out of ten companies was to increase productivity [23]. This objective was to be achieved without any loss in flexibility. Even though only a small portion of users, mostly large-scale manufacturers, introduce FMSs primarily with the intention of increasing flexibility, this objective has high priority among other users as well. In this context the FMS may also be considered as a *technology-oriented* concept to achieve flexibility – in contrast to flexible production islands with their organization-oriented approach.

As empirical studies on work organization and the integration of FMSs in the organization of production indicate, this concept is bound up with organizational structures characterized by a more or less strong division of labour between planning departments and the shop-floor. Basically, two factors can be used to explain this finding:

- FMSs demand a degree of investment in hardware, planning and implementation which far exceeds that required for the corresponding number of individual machine tools (whose share of the overall costs on average is below 50%). The profitability of such an investment presupposes that a very high degree of utilization (usually over 80%) will be attained. In order to achieve this target, the technical staff endeavours to reduce the influence of contingencies, carelessness, and maloperation, by automating the systems as extensively as possible (minimizing idle times by means of automatic planning, co-ordinating, and monitoring of the system internal procedures).

- FMSs are generally far from achieving the ideal of the complete manufacturing of workpieces. On average, only about 60% of the operations are carried out within the FMS [24]. For this reason, the co-ordinating of FMS operations with previous and following manufacturing steps is crucial. In view of management, this cannot be done by system operators without increased risks of mismatching and longer idle times.

The fact that a high degree of division of labour between the planning departments and the shop-floor is typical for most FMS cases cannot be seen as a result of technological determination. It is more the capital intensity and the sensitivity of the logistic FMS integration in job shop manufacturing which promote an office-oriented configuration and application. A few installations demonstrate that FMSs do not necessarily implicate a centralistic concept of organization [25,26].

In the context of the technical realization on the one hand and the scope for different forms of work organization on the other, an FMS, unlike systems used in planning and control like CAD/CAM or PPC, can hardly be classified into fundamental system alternatives. An FMS combines several components and control systems which may be evaluated according to whether, and to what extent, they promote or impede skill-based manufacturing. This has not only to refer to the "vertical" division between functions dealt within the FMS or other departments, but also to the division of labour within the FMS crew (Sect. 2.4.3). But, first of all, some empirical findings on development lines of FMS technology and work organization should be outlined.

2.4.2   Technical Concepts and Work Organization

For a long time the diffusion of FMS lagged far behind expectations and prognoses. Basically this was a result of the enormous planning, implementation, and investment costs, and the fact that it is hard to prove their economic advantages. A marked increase in diffusion in the 1980s was triggered by the growing supply of standardized components and systems. This increase was clearly based on the one- and two-machine systems which were the easiest to standardize. This was likewise the reason for the reverse effect: the stagnation in the application of large systems with more than five machine tools, which require an average of ten (!) man-years for conceptualization and implementation [27]. Despite these enormous efforts, there were and are many problems just in mastering the (software) problems of system control. Although some large systems with up to 33 machines have also been implemented, the "philosophy" of employing large systems has not yet gained wide acceptance [28]. Stand-alone, highly productive machining centres have proved to be more advantageous in practically all dimensions of flexibility (machining, set-up, retooling, etc.).

From a quantitative viewpoint, only one of the two basic FMS concepts has succeeded to date, namely the one consisting of machine tools of the same type which can *replace* each other. FMSs with different, *supplementary* machines designed to handle the complete machining of workpieces and therefore necessitating a larger system size have remained isolated solutions in individual companies [29]. The complexity of material flow organization (tools, workpieces) and the problem of differing machining times are two of the factors which so far have obstructed the availability of standardized FMS solutions involving various machining methods.

So far, however, the machining concept may have an influence on the degree of the "vertical" division of labour because planning and scheduling functions are easier to centralize or even automate in FMSs with machine tools of a single type.

The *size* of the system can have an impact on both dimensions of division of labour. On the whole, the larger the system and consequently, the larger the group working with it, the larger the scope for work organization in both directions:

- A single operator, responsible for a production cell or a two-machine system, is hardly able to assume planning and scheduling tasks to any considerable degree because of the time structure of necessary manual operations (machine-setting, chucking, handling breakdowns). This situation improves as soon as there are two or more workers in an FMS [30].

- On the other hand, the jobs in larger systems can be sub-divided to a greater extent than work at single NC or CNC machine tools.

Assuming that there are three basic alternatives of work organization as outlined above, the Tayloristic, the polarized, and the skill-based types (Fig. 2.6), hardly any of the systems investigated could be categorized into the last group. Even if there was a low degree of division of labour between system operators, most cases were characterized by relatively distinct dividing lines between the planning departments and system staff. For example, in none of the cases did programming belong to the regular duties of system operators, and planning and scheduling functions were assigned to the system staff (group leader) in a few systems only.

Overall there was a slight preponderance of FMSs with a high degree of division of labour between system operators (56% to 44% – see [23]). In this respect, hardly any deviations in terms of system size can be stated; except that two-machine systems were applied more frequently in set-ups with a high degree of division of labour. This is probably due to the fact that these systems are the most standardized and advanced with regard to their central computer and their integration into higher computer levels.

This refers directly to the impact of the *degree of automation*. The higher the degree of automation in the material and information flow as well as in monitoring, the greater the *potential for "decoupling"* the workers from

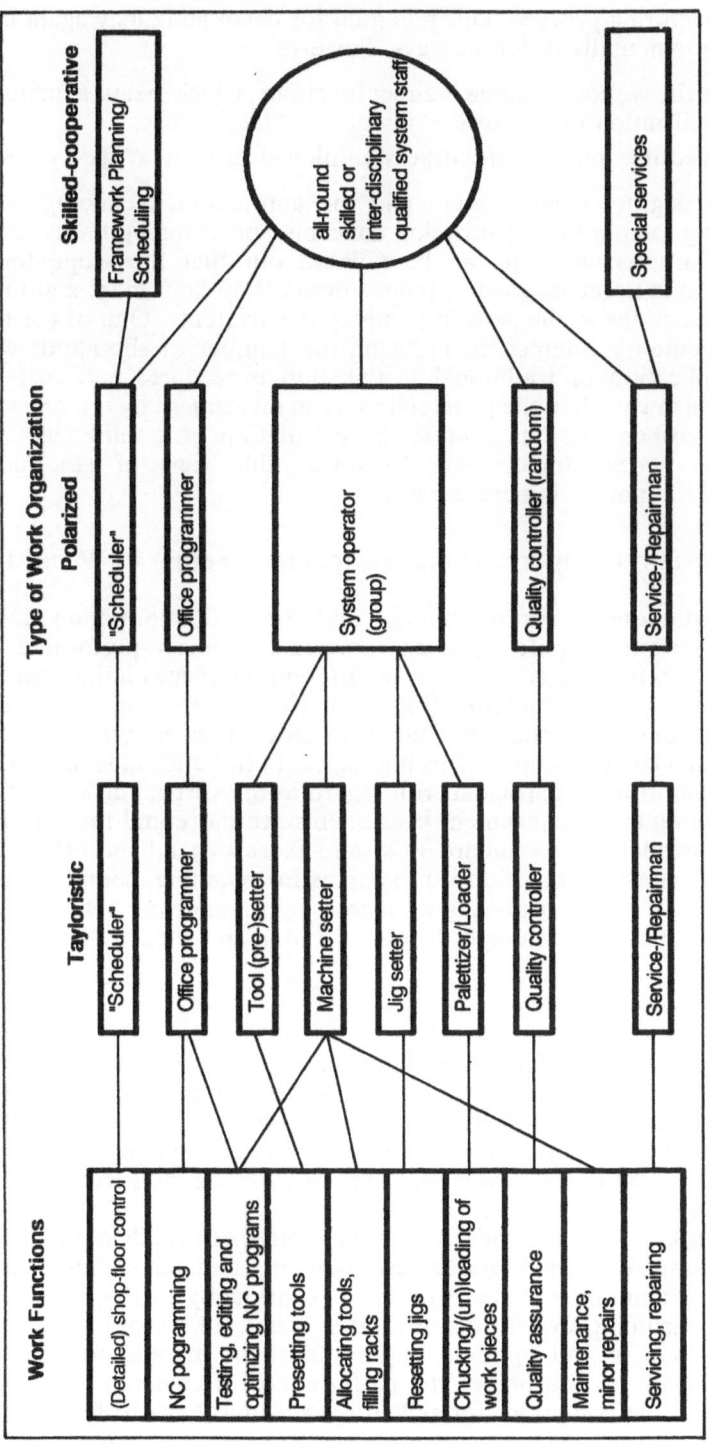

Fig. 2.6. Basic forms of work organization for FMS.

the manufacturing process. This potential for decoupling may again be used in two fundamentally different ways, however:

- To have the workers assume indirect functions, which means a qualitatively better utilization of workers' skills during "idle" times.
- To reduce the number of workers employed directly at the system.

Thus, the degree of automation and the automation technology are not determining factors for organization concepts and manning levels. But as a general characterization it can be pointed out that the scope for work organization is normally used to reduce direct shop-floor work and to assign planning functions to the staff in planning departments. One of the reasons for management's interests in reducing the number of shop-floor workers is the application of traditional justification procedures and cost–benefit analysis which calculate the profitability of an investment by the replacement of direct workers. In this context the calculation of a third "unmanned" shift is of special importance. However, this view of efficiency and profitability is not an imperative one.

2.4.3   System Components and Peripherals: Scope for Work Design

FMS subcomponents for machining, material and information flow take over and automate quite different functions, formerly performed by the workforce. Consequently, they have a different influence on the organization and design of production work (Fig. 2.7). Nevertheless, an examination of the system components does not suffice in order to assess the organizational potential and the limitations of technologies. Peripherals must also be taken into consideration, in particular with regard to the vertical division of labour since planning and management instruments are more and more frequently becoming available on standard PCs, and these expand the organizational scope quite considerably. Software systems for planning, configuration, and cost-evaluation are also playing an increasingly important role as technical factors influencing decisions in favour of FMSs and their design.

The Manufacturing System

The integration of different machining and processing methods offers a good opportunity to abandon the specialization of skilled work for certain machining methods and thus ensuring a flexible use of "interchangeable" workers. This suits, for example, the less specialized, young skilled workers trained according to the new training scheme for the industrial metal trades in Germany [31].

FMSs for large-volume and complex parts offer a basically more favourable time structure for skilled labour since planning and scheduling tasks, such as NC programming, can be carried out continually in the longer periods between machining cycles. On the other hand, FMSs designed for parts that need less machining time require frequent manual operations (i.e. chucking, loading) and hinder the performance of cognitive tasks. In such cases favourable conditions for skilled work would have to be intentionally created.

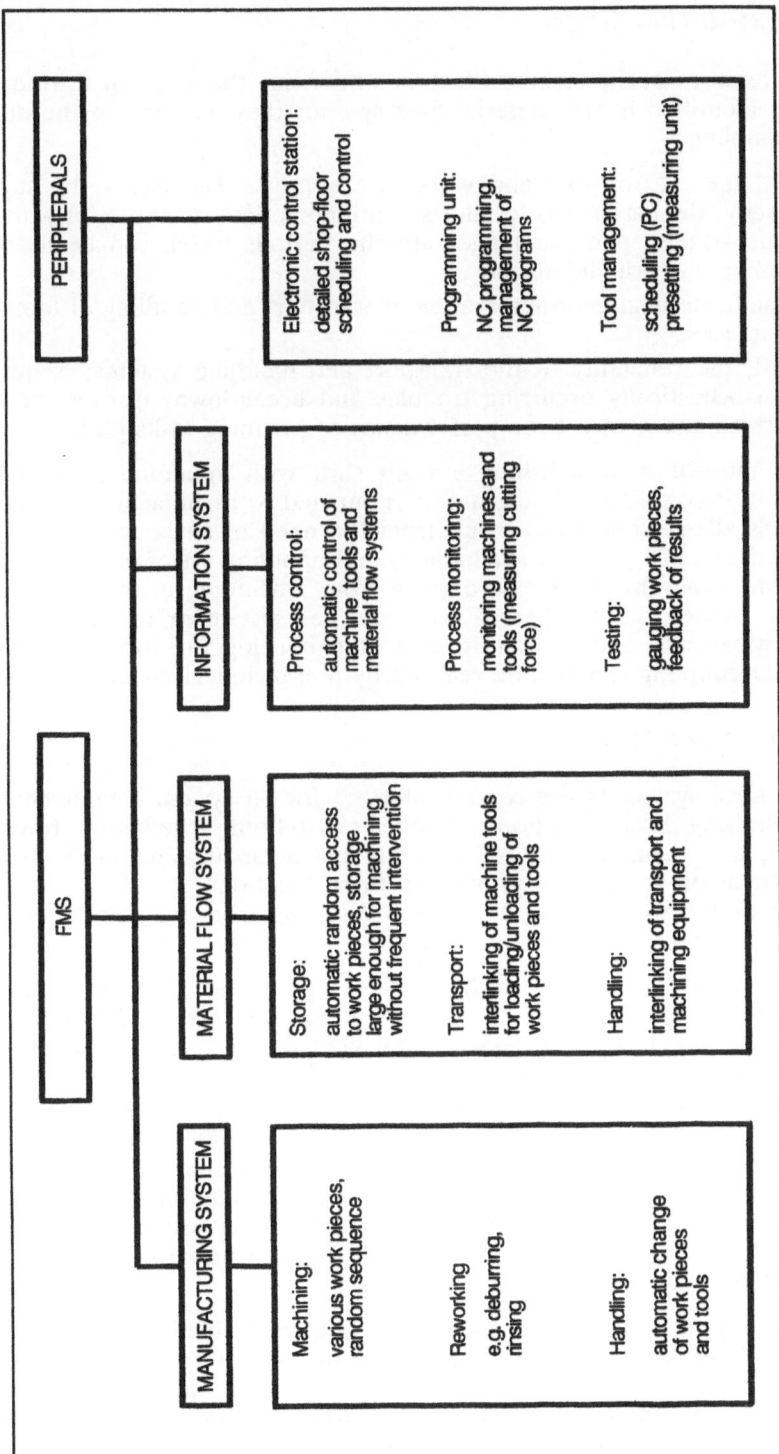

Fig. 2.7. Components and peripherals of FMS.

The Material Flow System

Three features of the technical factors influencing the scope in work design can be identified in the material flow system; they all concern the degree of decoupling:

- First, the size of tool and workpiece storage. The larger the storage capacity, the easier it is to allot a continuous time structure for activities at the system, and the longer are the periods which can be used for planning and scheduling jobs.
- Second, the degree of automation of transport and handling of tools and workpieces.
- Third, the reliability of the transport and handling systems; frequently and stochastically occurring troubles and breakdowns disrupt the time structure and obstruct the performance of planning tasks [32].

In the interest of establishing a night shift with minimum staff, a high degree of decoupling and automation is pursued by manufacturers and users of FMSs alike. Limitations result from the need of space for storage and the costs of tools (parallel availability) and handling systems.

On the other hand, the decoupling from machining cycles can also be used to reduce the number of operators per system or to assign two or more systems to one worker instead of enhancing the jobs by planning tasks. Decoupling can also be realized by non-technical means.

The Information System

The control system as the central interface for all system components has the following basic functions: detailed shop-floor scheduling, transport control, tool management, NC program management, and PDA. Not all the systems on the market provide all these functions.

With regard to order handling and *scheduling*, two strategic approaches can be distinguished, at least so far:

- First, control systems which provide algorithmic detailed shop-floor scheduling on the basis of internal allocation rules and optimization criteria, and organize necessary input via user guidance. Their objective is to convert centrally compiled job scheduling more or less automatically into an FMS-related procedure. This concept is frequently found in the first generation of standardized two-machine systems [30].
- The second line includes graphic interactive systems which are very similar to the ECS analysed in Sect. 2.2. Depending on the system, they offer a more or less large number of planning support functions. These systems are highly flexible and are suited for being employed by groups of skilled workers.

With regard to the *DNC function* as well, a distinction must be made between internal system tasks and the function of external integration. The internal system tasks (management and transfer of programs and corrected data) are to a certain extent an integral part of the operating system of the FMS and are significant for the work organization only insofar as they influence the degree of automation of the information flow and consequently the decoupling of workers from the internal processes.

The following tasks are the main ones performed externally with the DNC function: the link to an external programming system; the feedback of corrected programs to a centralized program management system; the transfer of machine and production data (running and idle times, number of workpieces, cutting speed, etc.). Here again, two typical concepts may be identified which differ in the extent to which they concentrate on office-oriented programming and a centralized compilation and processing of knowledge and experience, or whether they are open to decentralized forms of work organization (see Sect. 2.3).

Machine Control

Although these days every modern CNC control has the facility to enter data and edit programs manually, very clear differences may be identified in the programming concepts.

Computer-science-(and automation-)oriented concepts operate on the one hand with a higher "problem-oriented" programming language and aim at automatic programming (automatic calculation of geometry and technological parameters) for more complex parts. Systems on this basis require longer training periods and are usually not employed on the shop-floor. On the other hand, programming may be carried out semi-automatically or manually in a language close to the machine. Manual input in "low-level" language is very time consuming and therefore used infrequently, mainly for simple parts only.

Shop-floor-(and human-)oriented concepts can be categorized into three approaches:

1. The play-back method in which skilled workers create the program through a concrete machining process in a quasi-analogue manner and without any programming in advance. Despite its advantages – especially for experienced skilled workers – this approach has hardly been used as yet [33].
2. Systems with a dialogue interface based on clear text.
3. Graphic interactive programming systems which use descriptive symbols and graphic elements instead of a programming language. Using these, the workpiece geometry can be put together successively in some sort of a dialogue by means of soft keys. As a result, programming tends to be simplified and reduced to entering parameters.[6]

The basic approach of these concepts is to simplify and structure the programming in such a way that a skilled worker even without in-depth knowledge of programming is able to program the machines on the basis of his usual way of thinking and manner of working. This enables operators to become familiar with the systems quickly. The higher degree of

---

[6] Graphic-dynamic simulation makes the immediate visual control of the programming results possible, and is backed up with automated plausibility controls and collision monitoring.

programming supports guarantees high productivity in programming directly at the machine which means a high efficiency in learning and working. As a result, skilled workers are not only able to assume programming tasks but are encouraged to do so as well.

However, ISO-oriented controls are increasingly being equipped with appropriate programming tools as well. Even though producing a program may still remain more abstract, under certain circumstances programming complex parts can be more flexible. With a progressive, computer-science-oriented solution, the skilled worker is not submitted to a computer-guided dialogue. However, this would demand a greater amount of training and relatively higher expenditures for programming of parts that are not very complex [19].

Thus, between the purely office-oriented and the explicitly shop-floor-oriented programming systems, a range of control systems is establishing itself which makes computer-aided skilled work a feasible alternative.

Peripherals

Decreasing prices and constant improvements in efficiency on standard PCs are increasing the availability of every imaginable computerized aid to planning, scheduling, co-ordinating, and control of production processes at every level of a company. Technical limitations to reintegrating conceptual work on the shop-floor, which result from the installation of FMS components inadequate for this purpose, may almost always be compensated for today by supplementary FMS peripherals.

If, for example, machine controls or DNC computers are not suitable for shop-floor programming (e.g. because programming parallel to machining is not possible), programming units on a PC basis optimally adapted to the parts-spectrum may be used close to the system. Ideal for this type of application are systems which were developed in the "WOP"[7] project; they not only offer a uniform dialogue for all manufacturing methods but also a uniform procedure for office and shop-floor.

Inadequate user support in regard to detailed production planning (job sequencing, machine scheduling, etc.) by the FMS control station and the integration into the whole production process no longer constitute an argument against operating the FMS by the system staff itself since highly convenient and shop-floor-suited ECSs are available, and their investment costs are negligible compared with those of the entire FMS.

The same applies to the functions of tool presetting and tool management. Optical and opto-electronic presetting equipment can be installed directly at the FMS, or, in the case of insufficient utilization, in the centre of a larger shop-floor area. Easy-to-handle tool management computers are well able to support FMS external tool management by a group of skilled workers.

---

[7] "Shop-floor-oriented Programming Procedures" (in German: WOP) was the common heading of a government sponsored project under which a number of companies and engineering institutes co-operated.

2.4.4 "Decoupling" People by Technical Means is Not Sufficient

Even though in the individual case there might well be technical limitations to integrating jobs and forming skilled work groups on the shop-floor level, there is nevertheless a wide range of FMS components and peripherals available. Further flexibility and shop-floor-oriented systems appear to be current trends as well.

The possibility of using the existing scope of organizational design to promote computer-aided skilled production work has proved to be centrally dependent on non-technical parameters of *manpower requirements policy*. Whether the "decoupling" from the production and work processes may also become effective for system operators, and whether the higher degree of automation does in fact grant more autonomy to operators rather than just increasing the system pace, will finally be determined by the number of workers employed at the system.

A relatively too small number of workers operating the system means that the work of the system staff has to be reduced to palettizing or system (un)loading and setting of jigs and fixtures. Operating the system then usually means nothing more than some sort of multi-machine operation. This form of employment is encountered frequently because it corresponds to the attitude towards costs in the classic rationalization approach which utilizes automation-related time-saving to minimize shop-floor labour [22].

Even if setting-up and loading do not account for all the working time, the time structure of necessary job performance often prevents the execution of other and additional jobs. The remaining idle time between loading operations is fragmented by other instances requiring intervention, caused in particular by stochastic disturbances. These are mainly trivial disruptions lasting up to ten minutes (cf. [34], p 98). Due to these factors, sufficient time spans and structures for scheduling, programming, preventing breakdowns, and quality-related activities cannot be created by technical means alone; work organization and work design must create the necessary preconditions. The assignment of planning task to the system personnel is at the same time the prerequisite for scheduling the work procedures in such a way that sufficient intervention-free time is created, e.g. for programming.

With regard to the use of labour, Roth and Königs [35] could prove on the basis of an FMS design example applied in the automotive industry that a larger system staff not only makes skilled group work possible in the first place but that this is also profitable, particularly due to the higher degree of FMS utilization achieved and due to lower costs outside the system.

Manpower policy creates, at second glance, an enormous potential for organizational alternatives which are well able to compensate existing technical limitations. In this context a further category of computer-aided tools for the implementation and utilization of FMSs in companies should be mentioned: planning, configuration, and cost analysis systems for investment decisions regarding FMSs, which in the meantime are widely available and more and more are appearing on the market. These planning systems are gaining influence on the technical and organizational design of FMSs; however, so far they provide no means for judgements about the significance and efficiency of various forms of work organization, and the number and skills of staff. In this respect there is still a large gap in research and development.

This methodological deficit generally relates to the comparison of basic alternatives concerning technology and organization. Thus, bearing the scope for different design alternatives of FMSs in mind, there are many reasons for assuming that the less automation-oriented concepts of *flexible production islands* are better suited both to skilled and co-operative group work as well as to company interests in low-investment and thus low-risk increase of productivity and flexibility. To prove this, an adequate forecast of the cost-effectiveness would be desirable for companies on the brink of innovation. It should be based on an extended set of cost-relevant parameters and allow for considering important long-term, indirect and qualitative effects of organizational alternatives (ability for short throughput times and high reliability of delivery dates, costs of indirect labour and labour turnover, motivation of workers).

In the technical and organizational planning of FMSs or production islands, manpower policy based on preserving or enhancing the employment of skilled labour can be backed up by *prospective work analysis procedures*. These enable the evaluation and possible modification of alternatives beforehand in accordance with criteria – like skills and workload – before facts and constraints are created that are difficult to correct or reverse. Practical experience with methods of prospective work analysis and work design are already available [30,36].

2.5   Resumé: New Chances for Skill-Based Manufacturing?

In the early 1990s, rationalization in industry is characterized by an increasing diffusion of computer-aided and computer-integrated manufacturing technologies. At the same time, no clearly predominant concept of reorganizing and restructuring industrial work has emerged. To simplify matters, there are three different models of work on the shop-floor:

• Computer-aided "neo-Taylorism"
• Polarized production work
• Skilled and co-operative production work

While some companies are attempting to make traditional Tayloristic forms of work organization more efficient by utilizing more computer-based technologies, others are seeking ways and means of maintaining or revitalizing forms of skilled production work on the shop-floor.

According to practical experience to date, as well as in the opinion of many experts, only forms of work organization which are oriented to concepts of skill-based manufacturing will be able to offer sufficient chances not only for efficient and flexible production, but also for ensuring the availability of skilled labour for industry in the long run. The concept of skilled and co-operative production work would seem to benefit both companies and workers alike within the context of changing sales conditions and an increasing utilization of CIM technologies. This development option may therefore be regarded as a *guideline* for work organization for large sectors of the metal-working industry. As a result of a significant reduction of the so far mostly distinct division of labour in the dimensions of hierarchy, skills, and functions, this model seeks to create work structures which are characterized by a high level of skills of all workers, mutual interchangeability of tasks and jobs, as well as self-co-ordination and co-operation within work groups. However, whether or not this concept will be able to establish itself as a more or less general industrial practice depends on certain preconditions.

One major factor is the selection and utilization of technological systems which support forms of skilled production work or at least do not obstruct or prevent them. The relationship between technology and work organization which we have attempted to outline, will continue to play an important role in the ongoing and future changes in industrial work and structures of industrial labour.

A crucial prerequisite for skilled production work is the further development and utilization of computer systems which are flexible with respect to work organization and are open for adaptation to shop-floor conditions. These systems offer the largest possible scope for (re)-integrating conception and execution; this is particularly relevant in respect to CAD/CAM and PPC systems which are directly correlated with the existing hierarchical and functional division of labour in companies and thus with the associated separation or integration of conception and execution. Technology may be differentiated between two fundamentally different concepts:

1. Systems which entail the stabilization of or even an increase in the traditional forms of organization based on a high degree of division of labour.
2. Undetermined system concepts which offer a high degree of flexibility with regard to work organization, and by a specific design enable them to be utilized in the offices of planning or production engineering departments as well as on the shop-floor.

This chapter presents results of our analysis in respect of this dichotomy with regard to various CIM components and systems on the market; special consideration is given to FMSs, their components and peripherals, to manufacturing control systems and electronic control stations in particular, as well as to DNC and CAD-NC systems. In all these technology lines we find a more or less clear dominance of automation- or office-oriented concepts; at the same time there are obviously increasing efforts made to develop alternatives which are more open for skill-based shop-floor utilization.

While the various types of CIM components on the market and in use do not necessarily *determine* the actual forms of work organization and work design, they must nevertheless be regarded as important factors which may either facilitate or obstruct the establishment of skill-based manufacturing and production work.

The type of *company implementation process* of CIM technologies, the way in which innovations in regard to technology and organization are managed, is another factor which has a decisive effect on the chances of achieving skilled production work. Whereas mere reactive manpower policies mean severe risks for productivity as well as the quality of working life in the long run, a strategy which is geared to the option of skilled and co-operative production work involves long term manpower planning including further training policies. While an increasing number of industrial companies would seem to be taking initial steps in this direction, it cannot, at present, be seen as the mainstream of development in industrial labour.

References

1. Sauer D, Deiß M, Döhl V, Bieber D, Altmann N. Systemic rationalization and intercompany divisions of labor. In: Altmann N, Köhler C, Meil P (eds) Technology and work in German industry. Routledge, London, 1992
2. Schultz-Wild R, Nuber C, Rehberg F, Schmierl K. An der Schwelle zu CIM. Strategien, Verbreitung, Auswirkungen. RKW, Eschborn, and TÜV-Rheinland, Köln, 1989
3. Schultz-Wild R. Diffusion of CIM-Technologies. Dynamic dissemination and alternative paths of innovation. In: Altmann N, Köhler C, Meil P (eds) Technology and work in German industry. Routledge, London, 1992
4. Piore MJ, Sabel CF. The second industrial divide. Possibilities for prosperity. Basic Books, New York, 1984
5. Brödner P. Fabrik 2000. Alternative Entwicklungspfade in die Zukunft der Fabrik. Edition Sigma, Berlin, 1985
6. Warner M, Wobbe W, Brödner P. (eds) New technology and manufacturing management. Strategic choices for flexible production systems. John Wiley & Sons, Chichester, New York, Brisbane, Toronto, Singapore, 1990
7. Kern H, Schumann M. Das Ende der Arbeitsteilung? Rationalisierung in der industriellen produktion. CH Beck, München, 1984
8. Köhler C, Schmierl K. Technological innovation. Organizational conservatism? In: Altmann N, Köhler C, Meil P (eds) Technology and work in German industry. Routledge, London, 1992
9. Brödner P. Technocentric-anthropocentric approaches. Towards skill-based manufacturing. In: Warner M, Wobbe W, Brödner P (eds) New technology and manufacturing management. Strategic choices for flexible production systems. John Wiley & Sons, Chichester, New York, Brisbane, Toronto, Singapore, 1990
10. Hars A, Scheer AW. Entwicklungsstand von Leitständen. In: von Behr M, Köhler C (eds) Werkstattoffene CIM-Konzepte. Alternativen für CAD/CAM und Fertigungssteuerung. KfK-PFT 157, Karlsruhe, 1990, pp 51–78
11. Köhler C. Nutzungsformen elektronischer Leitstände. Ergebnisse einer Anbieterbefragung. In: von Behr M, Köhler C (eds) Werkstattoffene CIM-Konzepte. KfK-PFT 157, Karlsruhe, 1990, pp 79–100
12. Freidrichs P, Gromotka W. Fertigungsleitsysteme. Marktübersicht und Gruppierung. VDI-Z, 1989; 131 (8): 97–105
13. Herterich R, Zell M. Dezentrale Fertigungssteuerung. Neue Ansätze zur interaktiven Steuerung teilautonomer Bereiche bei Einzel- und Kleinserienfertigung. VDI-Z, 1989; 131 (5): 19–25
14. Manske F. Alternative strategies of production planning and control (PPC). In: Brödner P (ed) Strategic options for "new production systems". CHIM: Computer and Human Integrated Manufacturing. FAST Occasional Papers no. 150, EEC, Brussels, 1987, pp 131–140
15. Manske F. Neue Zeiten. Neue Formen der Kontrolle und Rationalisierung von Arbeit. Edition Sigma, Berlin, 1990
16. Strack M. Optimale Produktionssteuerung. Organisation, Wirtschaftlichkeit und Einführung konventioneller und EDV-gestützter Leitstände. TÜV Rheinland, Köln, 1986
17. Gottschalch H. Examples of human-centered work design in CIM-Structures. IFAC, Proceedings of the Symposium on Skill Based Automated Production, November 15–17, Vienna, 1989
18. Jackson S, Browne J. An interactive scheduler for production activity control. Int J Comput Integr Manuf, 1989; 2 (1): 2–14
19. Böhle F, Milkau B, Rose H. Computerized manufacturing and sensory perception. New demands on the analysis of work. In: Altmann N, Köhler C, Meil P. (eds) Technology and work in German industry. Routledge, London, 1992

20. Noble DF. Forces of production. A social history of industrial automation. Alfred A. Knopf, New York, 1984
21. Grupe U, Hamacher B. Werkstattorientierte Auslegung und Entwicklung von CAD-CAM-Systemen. In: ISF München (ed): Arbeitsorganisation bei rechnerintegrierter Produktion. KfK-PFT 137, Karlsruhe, 1988, pp 43–69
22. Martin JM. You can reduce manufacturing costs. Manuf Engng, 1988; 100 (6): 42–47
23. Fix-Sterz J, Lay G, Schultz-Wild R, Wengel J. Flexible manufacturing systems and cells in the Federal Republic of Germany. In: Warner M, Wobbe W, Brödner P. (eds) New technology and manufacturing management. Strategic choices for flexible production systems. John Wiley & Sons, Chichester, New York, Brisbane, Toronto, Singapore, 1990, pp 191–206
24. Eversheim W, Schönheit M. Kostenstrukturveränderungen flexibler Fertigung – Erste Phase bei der Entwicklung eines PC-Tool zur Kostenbewertung. VDI-Z, 1989; 131 (7): 64–68
25. Schultz-Wild R, Asendorf I, von Behr M, Köhler C, Lutz B, Nuber C: Flexible Fertigung und Industriearbeit. Die Einführung eines flexiblen Fertigungssystems in einem Maschinenbaubetrieb. Campus, Frankfurt, New York, 1986
26. von Behr M, Hirsch-Kreinsen H, Köhler C, Nuber C, Schultz-Wild R. Flexible manufacturing systems and work organization. In: Altmann N, Köhler C, Meil P (eds) Technology and work in German industry. Routledge, London, 1992
27. Shah R. Flexible Fertigungssysteme in Europa – Erfahrungen der Anwender. VDI-Z 1987; 129 (10): 15–22
28. Childs JJ. Japan show moves away from large-scale FMS. FMS Mag 1989; 7 (1): 41–43
29. Schönheit M, Wiegershaus U. Flexible Fertigung; Einsatzbedingungen von flexiblen Fertigungssystemen. VDI-Z 1990; 133 (1): 37–44.
30. Moldaschl M, Weber W. Prospektive Arbeitsplatzbewertung an flexiblen Fertigungssystemen. Psychologische Analyse von Arbeitsorganisation, Qualifikation und Belastung. TU Verlag, Berlin, 1986
31. Schultz-Wild R. Bringing skills back to the process. In: Altmann N, Köhler C, Meil P (eds) Technology and work in German industry. Routledge, London, 1992
32. Moldaschl M, Frauenarbeit oder Facharbeit? Montagerationalisierung in der Elektroindustrie II. Campus, Frankfurt, New York, 1991
33. Hirsch-Kreinsen H. On the history of NC-technology. Different paths of development. In: Altmann N, Köhler C, Meil P (eds) Technology and work in German industry. Routledge, London, 1992
34. Wiendahl HP, Springer G. Untersuchung des Betriebsverhaltens flexibler Fertigungssysteme. ZwF, 1986; 82 (2): 95–100
35. Roth S, Königs P. Gruppenarbeit als Gestaltungsalternative bei CIM-Einsatz. In: Roth S, Kohl H (eds) Perspektive: Gruppenarbeit. Bund, Köln 1988
36. Volpert W, Kötter W, Gohde, HE, Weber W. Psychological evaluation and design of work tasks: two examples. Ergonomics 1989; 32 (7): 881–890

3 Human-Centred Flexible Manufacturing Systems in Machining and Assembly

N. Mårtensson, L. Mårtensson and J. Stahre

3.1 Introduction

As soon as NC machine tools, industrial robots and automatically guided vehicles had proved themselves to be productive and reliable as separate units, the manufacturing industry started to combine them and various other equipment into systems aiming at flexible transfer line production. This happened in the late 1970s, although some efforts had already started ten years before, e.g. Sundstrand Machine Tool Company in the USA or Molins Machine Co. in the UK with the famous "System 24". In 1980 a German study [1] estimated the world-wide number of Flexible Manufacturing Systems (FMS) at about 80. Five years later a report from the United Nations Economic Commission for Europe reported the number to be 350 [2]. A fair estimate is that the world-wide number by 1992 ought to be between 1000 and 1500. Since there is no distinct definition of FMS, it is difficult to derive any accurate estimate. The figures, however, may indicate the volume of enterprises, machines and people involved.

The majority of FMS installed in industry are found in metal cutting operations. Other big application areas are welding and assembly. The latter ones are referred to as Flexible Assembly Systems (FAS). Some systems are very large, employing 20 machine tools or more, but the majority involve only five machines or less [2]. In metal cutting applications, FMS are often build around machining centres, which facilitates resetting and in general gives a reliable handling of tools and workpieces. Most FMS are tailor-made for the user, and the flexibility without considerable resetting is usually limited to a fairly narrow spectrum of variants of a few products. It was soon realized that implementing and running an FMS required close attention from qualified people if it was not to be a lost investment. Not only the technology but also organizational and managerial principles were changed. Much of the responsibility for the efficient daily operation of the FMS was shown to depend less on the technology and more on the persons attending it. This has created an interest in studying the human–machine–system relationship.

3.2   Flexible Manufacturing Systems (FMS)

3.2.1   FMS Technology

FMS is the organization of a manufacturing complex, by employing technology and people, into a production unit able to produce a spectrum of products at competitive prices and quality with short and precise lead times. The competition for FMS is found either in traditional labour-intensive manufacturing relying on peoples' skill and capacity, or in capital-intensive manufacturing relying on fixed automation equipment. Labour-intensive manufacturing is very flexible, but suffers from long and uncertain lead times, less consistent quality and, as wages rise, also from high cost. Fixed automation is very competitive on deliveries and quality consistency, but lacks flexibility and needs high production volumes to be cost-effective. The segment between these two extremes is increasing in importance, because of changes in market demands and customer preferences. That segment is the focus for FMS applications.

As stated in the definition, the FMS is a combination of organizational, technical and human efforts. The principle of the FMS organization is to obtain a straight flow of material through the manufacturing process, with one operation following directly upon another, with virtually no delay or buffering between. This is an efficient way of maintaining precise lead times and reduce capital tied up in work-in-process. However, it is also a physically vulnerable way of organizing any system. In this case it might be justified, due to the ease of planning and the strong need for short lead times. Success will depend highly on reliable performance of the processing equipment.

The technologies employed must care for high flexibility regarding product variation, and must therefore be of programmable type such as numerically controlled (NC) machine tools, industrial robots (IR), automatic guided vehicles (AGV), and computer aided design and manufacturing (CAD-CAM). These technologies have developed over the last 30 years into well-known and well-reputed components of manufacturing. Applied in FMS, however, they may require closer attention, since they will now work not as stand-alone units but in a system, depending on each other's performance. For example, NC machine tools work under geometrically well-defined conditions, both regarding the parts that are being cut and the workspace in which the tool is active. Also, the operation sequence is fixed and predetermined. It is a deterministic activity, lending itself well to off-line preprogramming. The industrial robot is less well defined geometrically. It is acting towards a periphery which is geometrically unstable. Unlike the stand-alone NC machine tool, the industrial robot must adjust geometrically to other machines and components for its operations, and even small positional errors will be hazardous. In addition, the robot's own kinematic behaviour lacks accuracy. The sequence of operations, however, can be predetermined and preprogrammed.

AGVs can be regarded as mobile robots in the above mentioned respects. They require a similar well-defined geometry in their workspace but execute a predetermined activity sequence.

CAD-CAM operates in the virtual world of information processing and has as such no physical restrictions. It is, however, highly dependent on the accuracy of the models used for actual products and machines.

3.2.2   FMS System Design

Designing and engineering an FMS is not only to pile together a number of reliable machine units. To work as a system towards a common goal, these units have to be interconnected. The interconnections may be mechanical, like feeders, fixtures, docking stations; or electrical, like switches, connectors, cables; or computer programs for signal processing and monitoring and machine communication. All these are often poorly specified, using partly unknown components at the time of design, and they may end up in the final system, unproven and with unknown endurance in continuous operation. This applies to software as well as hardware.

Developing the system will involve a number of specialists with different expertise. Usually each specialist assumes a non-changing, predetermined behaviour of the system. However, due to the numerous parts of the system, both software and hardware, and the interplay between their respective tolerances, the system is likely to deviate considerably from ideal behaviour. The overall result is that even a moderately complex system will stop frequently and, unless very competently operated by the experts who designed it, will give poor availability. The specialists, however, will leave the daily running to a non-specialist operator. He may be an experienced machine tool operator, but even so he will have difficulties reading and interpreting the signs as to whether the system's performance is good or bad. When a breakdown occurs he has little chance of identifying the source of the error, and can only wait for the specialist to come and do the repair. Apart from adding to the poor availability, this incapability of the operator is, of course, frustrating and highly demotivating.

3.2.3   FMS Operation

The FMS operation may be considered to consist of three major parts:

- Planning/scheduling the system according to a previously developed master plan, e.g. by MRP (manufacturing resources planning).
- Programming the equipment that constitutes the system.
- Dispatching/monitoring the system to fulfil the aims of the master plan.

Scheduling even a moderately complex FMS is not trivial. A real flexible system is constantly exposed to different operation sequences for different products or product variants in addition to changing priorities and other production conditions. The scheduling must in every case consider the availability and finite capacity of each resource of the FMS, as well as production due dates of all product batches being scheduled. Various techniques are employed in the scheduling process, most recently computer simulation of the production flow.

Programming the machine tools and other equipment of the system is only partly done off-line, since e.g. handling operations by industrial robots need to be adjusted on-line. However, all basic machine programming is performed prior to actual operation. Programming during operation is therefore done only to improve the original program or as reprogramming due to unforeseen events. Dispatching and monitoring has the main purpose of reacting to the current state of the system and selecting the best possible action. Dispatching is responsible for dealing with unplanned occurrences, and in order to do so, it needs dynamic manufacturing information from the monitoring function, describing the current status of the system. Handling unexpected events is thus a key function in FMS operation. Due to the complexity of the task, it is a job for the human operator. The man–machine interaction in this situation is a main target for human-oriented FMS design.

To sort out what is required from the system of man and machine tools it is convenient to identify the task in FMS operation as related to machine functions and operator behaviour.

3.3   FMS: Technology Tasks

A manufacturing task has an ideal specification where all task parameters are given set values. This constitutes the base for the task model which is used for the programming. The deviations of the parameters from the set values, due to various physical deficiencies in the system, will define the real task. It is practical to consider three levels of deviation.

Deviation Within a Specified and Accepted Tolerance.   This first level defines by tradition the ideal manufacturing situation. A typical first level task could be a shaft-turning operation, where all machines and parts are within specifications. As stated before, all geometric parameters and the operation sequence are within predetermined and accepted tolerances. This ideal deterministic manufacturing situation is completely predictable and programmable.

The deviations are in this case dealt with by using machines or external devices that are built to absorb the accepted tolerances. For instance, machine tools define the geometry of a workpiece by a reference cut. Industrial robots use precise fixturing.

Deviation Outside Specified Tolerances, but of a Known Character.   The second level is characteristic of a general manufacturing situation when machines may malperform slightly or accidentally be off position, and parts may not be to specification. As mentioned before, an industrial robot falls within this category when performing a close-fitting insertion in mechanical assembly. The distribution of tolerances of the assembled parts may combine into a play between parts that is much smaller than the repetitivity of the robot. This is an example of geometrical constraints difficult to handle automatically. As a side effect it also illustrates the difference, from a manufacturing point of view, between tolerance, which absorbs predictable process deviations, and play where the absolute value is unpredictable.

Even more significant is the random sequence of operations. Because of the flexibility of the FMS regarding product and part variation, the planning may frequently change the operation sequence, depending on the status of the resources of the system, in order to satisfy the overall goal strategy.

This extension of the first level situation is extremely relevant to integration in manufacturing systems. FMS tasks are therefore by nature on this level of uncertainty. Even a small system or a cell will be complex enough for a non-deterministic behaviour. Left to itself, the second level of deviations may cause anything from faulty parts to a major machine breakdown. However, the type and range of deviation in each cycle of operation is predictable and strategies for recovery to an acceptable state can be developed, namely machine monitoring, error diagnosing and proper recovery actions. To some extent, for a limited number of frequent and well-defined deviations, this can be performed by the use of sensor feedback. Research work is in progress in order to increase the application of sensor-based monitoring and control. However, the technology is not sufficiently developed to handle the full complexity of an FMS, but it is still important as support to human intervention.

Unpredictable, Unknown Deviations. This third level is not supposed to occur in manufacturing, but is generally characteristic of our daily life environment. The situation needs scene interpretation and mapping, and continuous revising and updating of the plan of action. An FMS with an unlimited flexibility regarding product variation may fall within this category of operation, but this is a far from realistic situation. This level of deviations needs not only sensor feedback but also a qualified rule-based strategy to plan and execute actions according to the schedule or master plan. In an unpredictable environment this is a formidable task, which cannot so far be handled without human intervention.

3.4 FMS: Human Tasks

3.4.1 Human Needs and Resources

FMS operation may thus be considered non-deterministic, dealing with activities of a known character whose geometry and sequence may be out of accepted tolerances. As stated, this situation is too complex to handle fully automatically in an effective way. The responsibility of the operation will thus fall on the FMS operator. Since he is a non-specialist, it is important that the system reacts towards him in a way that supports his level of knowledge and that motivates him to take action. Demands on the system are in this respect:

• System status presentation relevant to the situation
• Decision support to select best possible action
• Interactive system/operator interface, easy to manage

Development of these system functions is highly dependent on a correct model for the roles, abilities and motives of the operator in charge.

During the 1950s, Maslow [3] launched his theory about the human needs that grow in accordance with the development of the society. The physiological needs are the basic ones while the needs for social security, belongingness, self-development and knowledge for its own sake can be found higher up on the "stairs". Maslow's theory is still valid, although amendments have been suggested. Rubenowitz [4] states that "waves" is a more adequate description of the needs than "stairs", since the physiological needs are always present.

During the 1970s the German psychologists Hacker and Volpert developed a theory of action control as a basis for analysing the learning situation at work as reported by Aronsson [5]. Their work as well as Maslow's theory has influenced a group of Swedish researchers in their analysis of the qualitative work content in FMS [6]. Edgren et al. [7] discuss the resources and needs of the human being in the following way:

> Human beings have a real and an experienced capacity at the three resource-levels: the creative, the perceptual-cognitive and the sensory-motoric. If this capacity is not being used at work, it becomes a need within the individual, a need that is being satisfied outside work instead of at the work place.

The lowest resource-level, the *sensory-motoric*, comprises the activities where we use our physical means as hands, arms, shoulders, legs etc. This level also includes our sensory system such as sight, hearing, touching, taste and smell.

The intermediary level, the *perceptual-cognitive*, comprises the ability to receive information and to understand, learn and logically deal with the information from the base created by earlier acquired knowledge and experiences.

The highest level, the *creative*, is the development of new solutions based on the knowledge and experience obtained through the intermediary level, which in turn has been partly built up by means of the lowest level. Thus, the creative level stands for development.

The attempt to classify the work tasks at FMS with evaluations according to the need model is preliminary, and a first endeavour to measure the qualitative contents of work. Four case studies on FMS have been carried out in the research programme. Attention has been focused on the working tasks for operators both from a quantitative and from a qualitative viewpoint.

3.4.2 Quantitative Analysis of FMS and FAS Tasks

At the FMS installations for *machining* that were analysed, the operators carry out loading/unloading of workpieces, tool handling, maintenance, supervision, repair work, testing and correction of the programs and some part of the planning.

At an automated *assembly* station under investigation three robots carry out assembly followed by an automated testing equipment and an automated packaging equipment. As this equipment, however, could not manage all sizes of the product, manual assembly and manual testing also took place. The operator team consisted of 11 people working in two shifts. The team members worked according to a job rotation scheme with four workstations: supervising two robots, one robot and the testing and packaging equipment or doing manual assembly and manual testing. Table 3.1 shows the time spent per shift on each of the work tasks in the three companies with FMS and at the automated assembly system. It can be seen that the operator spends a considerable part of his/her 8 h shift in loading/unloading at the manufacturing system and in material supply at the assembly line. Maintenance of the system demands from 7% to 20% of the working-day and repair work takes 3%–8% at the manufacturing systems. Here the more complicated repair work is carried out by a repair man. At the assembly system the operators spend 27% of their time in repairing the robots. The table also shows that the programming work and the planning are minor tasks of the whole job. An important observation is the existence of unspecified time for the operators. This is time remaining after the obligatory tasks have been carried out. This time could be used for making improvements on the equipment, for communication with other departments and, perhaps, with suppliers of goods, for learning new tasks, for teaching newcomers etc. Altogether, the unspecified time could be seen as a prerequisite for a well functioning technical system and for skill development of the operators.

The quantitative analysis has given a picture of what the operators are actually doing at flexible manufacturing and assembly systems. It is, however, important to make a qualitative analysis of the work content as well.

3.4.3 Qualitative Analysis of FAS Tasks

The evaluations have been made on the earlier mentioned four case studies. The robot assembly station will be given as an example in Table 3.2. According to the table, the tasks that take most of the time are repair work, maintenance, supervision and supply of material. The evaluations of the qualitative content of the tasks are made in such a way that the total content of a task is marked 100 and a distribution of the total content is made on the three levels with the answer expressed with 5% precision. It shows, for instance, that material supply is up to 80% sensory-motoric and

Table 3.1. Work tasks at three flexible manufacturing systems (FMS) and one automated assembly system. Time spent on each task (% of 8 h shift) [6]

Work task	FMS			Assembly system
	Company A	Company B	Company C	
Loading/unloading	25	46	22	
Material supply				15
Tool handling	25	4	2	2
Maintenance	13	15	6	20
Supervision	4	(100)[a]	6	20
Disturbances/repair	8	3	4	27
Test, corrections of programs	8	4	9	3
Planning	4	6	9	6
Unspecified time	13	22	42	7
Total	100%	100%	100%	100%

[a] At company B it was stated that supervision takes place constantly.

Table 3.2. Work tasks at the "two-robot station", time spent on each task as a percentage of 8 h shift and quality of job content

Work task	Time (%)	Job content (%)		
		Sensory-motoric	Perceptual-cognitive	Creative
Material supply	15	80	20	0
Set up	2	40	50	10
Maintenance	20	50	40	10
Supervision	20	10	80	10
Repair work	27	30	50	20
Programming	3	5	55	40
Planning	6	10	55	35
Unspecified time	7	–	–	–

has no creative content at all. The planning, however, has been estimated to comprise 10% sensory-motoric, 55% perceptual-cognitive and 35% creative tasks; 6% of the working hours is spent on planning.

By multiplying each of the three qualities by the corresponding span of time for every task and adding them all up, the complete work content is obtained. Consequently, the work-day may comprise 15% creative, 45% perceptual-cognitive and 40% sensory-motoric tasks for the operators at the robot station.

Creativity in the work can be found in repair work, programming and in planning. The same estimations have been made for the manufacturing systems. However, programming and planning seems to occupy more of the time in machining cells than in assembly ones, while supervision and repair work is the other way around.

3.4.4   Conclusions of the FMS Analysis

Among the main conclusions regarding job content to be drawn from the above case studies of FMS and FAS are:

- Operators have many different work tasks.
- An FMS is a complex installation – a lot of disturbances may appear.
- An FMS does not work without an operator, who recognizes and deals with the disturbances and, even better, prevents them from appearing.

3.5   Operators Controlling the Machines

3.5.1   The UMIST Project [8]

Early work on human orientation of FMS was focused on the machines of the system and directed towards planning and programming tasks that allowed the operator primarily to make use of his professional skill.

Considerable effort was put into developing machine tools that were programmable by the operator and included some limited planning functions.

A pioneering research effort was carried out at University of Manchester Institute of Science and Technology (UMIST) during 1982–85 by Howard Rosenbrock and a group of engineers and social scientists. The title of the project was "A Flexible Manufacturing System in which Operators are not Subordinate to Machines". The objectives were to develop software which will enable the operator to program an FMS by making the first of a batch of parts, and in doing so, to develop a methodology for the simultaneous consideration of social and technical aspects during the development of new technology. The first part of the objective, the development of the software and the progress towards the technical solution was eventually defined as building an interactive CNC-control system in such a way as to give freedom to the operator to apply his machining skill as he seemed fit. This goal was reached, at least conceptually.

Underlying the goal definition was the view that NC methodology often assumes a deterministic machining process able to be planned and programmed away from the machine, level one in the previous task description. All implementation problems are identified and corrected during the programming and test running. For unforeseen, unpredictable events the operator has to intervene. Therefore this was regarded as an example of a technology that seeks to replace a machinist's skills, but still relies on the same skills to overcome its own deficiencies.

The approach taken for the design of the control system was to present the operator with a "blank table" where he could specify what he deemed as adequate machining parameters. The computer would then calculate additional, unspecified parameters and also constraints, should there be any, for the parameter values. In this way scientific knowledge expanded practical knowledge and no attempt was made by one to rule the other. The idea of the engineers was that the operator in this way would make optimal use of the computer and could spend time for more important tasks. The social scientists, however, meant that if the computer first gave a set of values, the operator would consider these values to be correct and not try to change them, in which case the operator's skill would not be developed.

The software was developed on a central computer, and was later to be rewritten to run on a microprocessor in the NC-control for the lathe. Colour graphics was used for displaying tables and bars. The programming was performed in Pascal. Declarative programming languages like LISP or PROLOG were not used, since they were considered insufficiently developed at that time. The stage of rewriting for the microprocessor was not reached during the project, but was carried over to the ESPRIT project 1217 "Human-Centred CIM Systems", which followed the UMIST project. This is referred to later on.

In the second part of the project, the simultaneous consideration of social and technical aspects in the design groups, the engineers and the social scientists had different views. The engineers approached the "human-centred design" task specifically from a wish to counter the effect of Taylorism, with which they were familiar, and this led them to emphasize autonomy and valuing the operator's skills, particularly tacit skills. The social scientists started, not from Taylorism specifically but from human development needs in general, and that included the importance of learning as well as other factors. This and other differences of opinion emanating from different backgrounds for engineers and social scientists came to occupy much of the project time. These discussions, however, were later considered a very valuable outcome of the project as they pointed to the different "cultures" and the necessity to reach consensus on goals and methods.

3.5.2   ESPRIT – Human-Centred CIM [9]

The original technical ideas of the UMIST project were expanded in the ESPRIT project 1217, Human-Centred CIM, which was to produce human-centred CIM components, in which human skill and its application are optimized in harmony with leading-edge computerized manufacturing technology.

The ESPRIT project has resulted in CAD, CAP and CAM products demonstrated at industrial sites in the three participating countries: Denmark, Germany and the UK [10]. The British project partners continued the development of CNC control for lathes. It is laid out as graphically interactive, and the processing operations can be simulated on a screen. The command language is directly oriented towards the geometrical and technical conditions of metal processing with machine tools which are either known to the operator, or are immediately understandable. No artificial information or technical terminology and no abbreviated command codes are transferred into the workshop. This control enables a program input directly at the machine. The quality of the CNC program is based on the knowledge and experience of the skilled worker.

Embedded in a CIM system as a CAM component, this new activity-oriented control is human-centred in that it permits technological competence to remain within the shop-floor and gives it support, instead of withdrawing this competence and transferring it to an NC-programming office.

The Danish partners in the project developed an electronic sketch-board, a portable digitizing board (20 × 30 cm) which can be operated in combination with a PC AT or the corresponding laptop. It is furnished with interfaces to the CAD system as well as to the production planning and control system. The sketch-board contributes to the design of human-centred CIM systems insofar as human communication is supported by information technology and because practical manufacturing experience has found a means of being integrated into the system.

The German partners developed a concept for extracting the planning and monitoring as well as the control of the order processing from the separate departments as far as possible, in order to link them with manufacturing work again. The organizational way to achieve this is by use of an autonomous production island. This island is supposed to perform tasks of operations planning, order processing and control together with the actual manufacturing in its own responsibility.

The research is being continued in several ESPRIT CIM projects as reported by Kidd [11], all of them carried out in interdisciplinary teams of engineers, computer scientists and social scientists. These projects aim at developing skill-supporting rather than skill-substituting technologies.

3.5.3 The Supervisory Control Model

The use of human behaviour levels in industrial applications has been described by Sheridan [12]. He suggests that useful operator decision support should be designed in a way that the operator

> downloads some of the rule-based and almost all of the skill-based commands (or programs) into the supporting computer, while retaining the knowledge-based tasks for himself.

To describe further the operator–process interaction, Sheridan has designed a supervisory control model. The model emerged from aircraft and spacecraft research in the 1960s and has been tested in continuous process industry. The supervisory control model (Fig. 3.1) is based on the assumption that a certain, semi-automatic process can be controlled and supervised through the use of artificial sensors and effectors. The human supervisor, or operator, receives information and gives commands to a human-interactive computer. This computer communicates with a process-interactive computer connected to the controlled process through sensors and effectors. Thus a two-way communication channel is established between the supervisor and the process. The model includes the definition of five roles for the human supervisory controller and their interrelated feedback loops:

- *Planning* is the first role and includes gaining enough understanding of the controlled process to be able to define a resource allocation strategy. The plan is dynamic in that it might be revised, should problems occur or the understanding of the process change. Plans should, if possible, be tested before being transferred to the teaching/programming phase.
- *Teaching* done by the supervisor means communicating the necessary commands, for the supervised system to act according to plan, to the system. Teaching is in this context analogous to computer system programming. The methods of programming might vary from pushing a

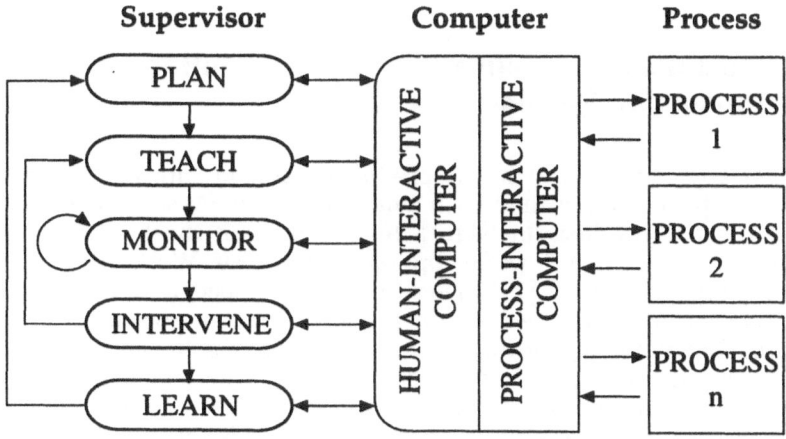

Fig. 3.1. Supervisory control model [12].

button or "teach and play-back" programming to abstract, high-level command languages, e.g. programming of industrial robots.

- When *monitoring* the operator observes the automatic execution of preplanned and preprogrammed system actions. The role forms a closed control loop by itself and is a normal state of an automatic system running without problems. Monitoring includes the acquisition of process status data and the estimation of acquired data. Finally the operator can evaluate the system status, drawing conclusions from acquired data and previous knowledge in relation to the desired system goals. The evaluation takes the form of a system diagnosis, which in the case of a failure is the basis for operative intervention in the system.

- *Intervening* in the process is conditional, and requires a failure or a system halt. If the system has stopped due to a known fault or to normal termination of the task, the operator should take action simply by starting the next process step. This generally means looping back directly to the teaching/programming role where a new process can be started. If the termination or failure is of an unknown type the operator will behave differently. Instead of looping directly to the programming role the new situation requires the operator to learn new conditions for a system fault.

- The *learning* process is entered every time an unknown error occurs, or when a normally terminated task has to be followed by the introduction of a brand new process. An operator should record the immediate events surrounding the unknown error, thus accumulating experience not only in the operator's mental model but in computer records available for future operators. The analysis of cumulative experiences with the controlled process closes a peripheral control loop back to replanning of system behaviour due to the supervisor's change in process understanding.

Sheridan developed his model aiming at conditions in the continuous processing industry, like power plants and chemical process complexes. An effort to study the applicability of the model in discrete parts manufacture, like FMS, is currently in progress [13, 14].

3.6 Operators Controlling the System – Anthropocentric Production Systems

Another approach to man in the production system with more or less the same basic philosophy as the one in human-centred CIM started in Germany. In 1983, Peter Brödner introduced his ideas about the benefits of group technology at an IFAC workshop on "Design of Work in Automated Manufacturing Systems" [15]. These ideas developed into the concept of anthropocentric technology. Brödner sees the anthropocentric quality of any technology as hinging on the way in which it is linked to the human faculties brought into play in the labour process. These faculties, which are unique to humans and not automatable, make for a smooth running of the work process. The preservation of these faculties is the supreme object of anthropocentric technology design. According to AT & S Newsletter [16]:

A technical system designed for use by humans can be called anthropocentric, if it facilitates human work processes by its very use and in conjunction with the working conditions.

APS balances technology, organization and human factors in a holistic concept. In more specific terms the anthropocentric production system is composed of four dimensions: education, management, organization and technology. These are applied on four levels: plant, interdepartmental relations, working group and workplace. Education and training are emphasized on the working group and workplace levels, in order to ensure competent personnel. Management is challenged with handling the process of change, in particular top and middle management. Organization and technology cover all levels with more or less equal importance (see Tables 3.3 and 3.4) [17].

Referring to the case of FMS, central organizational issues (Table 3.3) are low division of labour and integration of job tasks, like programming, scheduling, maintenance and processing. Semi-autonomous production islands may be grouped around part families. Co-operative relationships between white collar and blue collar workers are established. Central technology issues (Table 3.3) are the use of software to support APS structures and the relative unimportance of hardware components. Some examples of existing technology are the earlier mentioned products resulting from the ESPRIT 1217 project: the graphically interactive NC controller, the electronic sketch-pad and the group supporting scheduling system.

As a whole the technical system has to:

- Allow for a wide scope of action.
- Provide an adequate time scope.

Table 3.3. Anthropocentric production systems: organization design on different plant levels

Levels	Measures and Principles
Factory	• Small decentralized production units • Product shops • Companies within a company • Delegation of responsibility to lower levels
Inter-departmental relations	• Co-operation between design and manufacturing • Interactivity between workshop and engineering departments concerning programming of machine tools • Integration of business department, technical department and shop-floor, concerning planning and scheduling
Group	• Installation of production islands • Group work in FMS • Semi-autonomous assembly groups
Workplace	• Workshop programming • Integration of intellectual and manual functions • In highly automated areas: integration of programming, planning, maintenance and processing tasks • In low automated assembly areas: work enrichment with decision space on execution sequence and performance

Table 3.4. Anthropocentric production system: technology

Level	Technology
Workplace	• Shop-floor programming systems for machine tools and robots • Decision support systems • Analogue user support mechanism to control the manufacturing process • Symbolic representations of complete pictures for information, process and decisions • Skill supporting and learning techniques
Group	• Scheduling and planning systems for group work • Computer-aided co-operative work techniques for information, planning and decisions
Inter-departmental	• Information technology systems to facilitate interactions and dialogue between office and shop-floor • Transportable analogue design sketch-pads
Factory	• Information systems to support network organizational structures
General	• Adaptable and natural language human/computer interfaces • Highly transparent support systems for collective and individual decision making • New vision and symbolic representation systems

- Facilitate a mental grasp of and a personal strategy of coping with the work situation.
- Avoid objective obstacles to the work activities.
- Provide sufficient and varied physical activities.
- Call for the use of a wide variety of human sensory faculties.
- Provide for the tangible handling of objects.
- Allow for the shaping of working conditions according to individual needs.

The idea of anthropocentric production systems is now an object for research in the European Community within the FAST programme (Forecasting and Assessment in Science and Technology), and it is perceived as a productive and competitive tool for industrial modernization. Lehner [18] reports on the European response to Anthropocentric Production Systems. He finds that indicators of APS, such as shop-floor programming, group production concepts, multiskilling and interdepartmental collaboration have reached the stage of moderate to wide experimentation in Germany and in the Nordic countries, while developments in France, Belgium and the UK are limited. Ireland and southern Europe member states exhibit a few early approaches.

3.7 Discussion

Over the 15 or more years that an active pursuit of humanization efforts in manufacturing systems has been underway, there is a clear tendency to focus on ever more complex and complete systems. The preliminary studies of Rosenbrock et al. were concerned mainly with the interaction between man and a single machine. They were technology oriented and dealt primarily with programming and planning of a limited set of operations. One might say that they dealt with the deterministic first task level of the FMS, as it was defined in Sect. 3.3: "FMS: Technology Tasks". These studies were expanded in the ESPRIT project 1217 to more complex planning and monitoring situations, similar to the second task level. This is also the focus of the studies based on the Sheridan model, where, however, operator skill development and learning by using the system is emphasized.

The APS discussion in Germany and the European community is addressing a complete factory production system. The ideal system is seen less technology-oriented, with more emphasis on organizational, educational and managerial aspects. Here once again the problem appears of co-ordination and co-operation in research. Different disciplines and cultures must reach consensus on goals and methods, to be able to specify a comprehensive system design.

It is clear that the modern manufacturing system has a more complex function than was originally anticipated and that this function cannot be performed totally automatically. On the contrary, it turns out to be a paradox that the further automated the system gets, the more dependent it will be on human intervention to keep its functionality. The demands on the organization of the production, and on the people in this organization, however, are new and to some extent unknown. The pioneer work referred to in this chapter on the operator support in FMS is a first and significant contribution to the human orientation of future production systems.

References

1. Warnecke HJ, Kampa H. Flexible production systems in Western Europe – state and trends. In: Proceedings of the 4th International Conference on Production Engineering, Tokyo, 1980
2. Economic Commission for Europe in Geneva. Recent trends in flexible manufacturing. ECE/ENG.AUT/22, United Nations, New York, 1986. (ISBN 92-1-116347-1)
3. Maslow A. Motivation and personality. Harper, New York, 1954
4. Rubenowitz S. Organizational psychology and management. Esselte Studium, Gothenburg, 1987
5. Aronsson G. Demands at work and human development. Prisma, Stockholm, 1983
6. Mårtensson L. Flexible man in automated manufacturing and assembly systems. In: Proceedings of the IFAC world congress, Tallinn, Estonia, 13–17 August 1990, vol 6. Pergamon Press, Oxford, 1991
7. Edgren B, Islo H, Johansson G. Computer modelling of manufacturing systems. ISRN KTH/AVF/FR-91/4-SE, TRITA-AVF 1991: 4, The Royal Institute of Technology, Stockholm, 1991
8. Rosenbrock HH (ed), Designing human-centred technology – a cross-disciplinary project in computer-aided manufacturing. Springer-Verlag, London, 1989
9. Slatter R, Husband T, Besant et al. A human-centred approach to the design of advanced manufacturing systems. Ann CIRP Manuf Technol 1989; 38 (1): 461–464
10. Gottschalch H. Examples of human centred work design in CIM structures. In: Proceedings of the IFAC world congress in Vienna 1989, Pergamon Press, Oxford, 1990
11. Kidd PT. Research policy review. Organization, people and technology in European manufacturing: interdisciplinary research for the 1990s. Int J Hum Factors Manuf 1991; 1 (3): 257–279

12. Sheridan TB. Supervisory control. In: Salvendy, G. (ed) Handbook of human factors. John Wiley, New York, 1987, pp 1243–1263
13. Mårtensson L, Stahre J. Operator roles in advanced manufacturing. In: Brödner P, Karwowski W. (ed) Ergonomics of hybrid automated systems III. Elsevier, Amsterdam, 1992, pp 155–162
14. Stahre J. Humanufacturing – operator decision support in advanced manufacturing systems. Chalmers University of Technology, Göteborg, 1992
15. Brödner P. Group technology – a strategy towards higher quality of working life. In: IFAC workshop, 7–9 November 1983, Karlsruhe, Federal Republic of Germany. Preprints VDI/VDE-Gesellschaft Mess- und Regelungstechnik, Germany, 1983
16. Anthropocentric Technology & Systems (AT & S) Newsletter no. 1. Monitor/FAST Programme, Brussels, Feb 1990
17. Wobbe W (ed). Anthropocentric production systems – a strategic issue for Europe. Commission of the European Communities, Brussels, 1991 (FAST/APS research papers series, vol 1)
18. Lehner F. Anthropocentric production systems – the European response to advanced manufacturing and globalisation. Commission of the European Communities, Brussels, 1991 (FAST/APS research paper series, vol 4)

4 Unskilled Worker-Oriented Manufacturing

J. Iimura

The decline in the working population is making it more difficult to preserve the work force. If the decline continues, it will be difficult for the manufacturing industry to maintain the current level of production. With this in mind we must consider the roles of both skilled and unskilled workers in production processes.

4.1 Roles of Skilled Workers in Production Processes

In general, people in industrialized countries tend to regard manufacturing as dirty, demeaning and labour-intensive. This makes the declining working population a more serious problem for the manufacturing industry than for other industries, such as the service industry. As a result, the number of unskilled workers in manufacturing is increasing, and there is an urgent demand for computer systems which will allow unskilled workers to maintain high performance levels. In addition, highly computerized manufacturing requires highly qualified workers who can operate systems effectively and efficiently. Based on the evidence for the working population in Fig. 4.1, some basic and related knowledge on unskilled worker-oriented manufacturing is described in the following. Figure 4.1 shows the transition of the working population in Japan.

4.1.1 Classification of Jobs of Skilled Workers

When considering a computer system designed for unskilled workers, it is useful to classify the jobs of skilled workers according to the work processes involved. Each classified process may require a different computer system. Also, creating a model system helps clarify the flow of products and information. There are various methods of classifying work processes. The classification system used in this chapter is based on eight work processes which are described in the chronological sequence in which they occur: production planning, part procurement, selection, work instruction, flexible manufacturing systems (FMS) and just-in-time (JIT), information collection and abnormality detection, and judgement and improvement. Figure 4.2 illustrates classification according to these chronological processes.

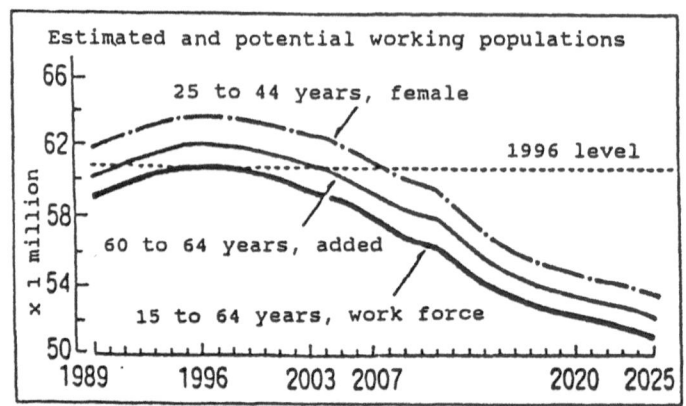

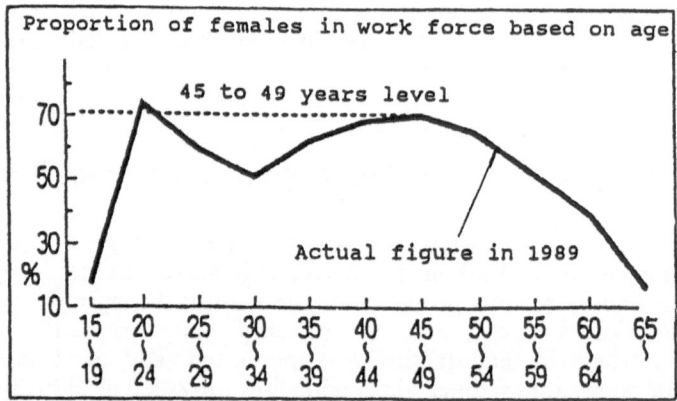

Fig. 4.1. Transition of working population.

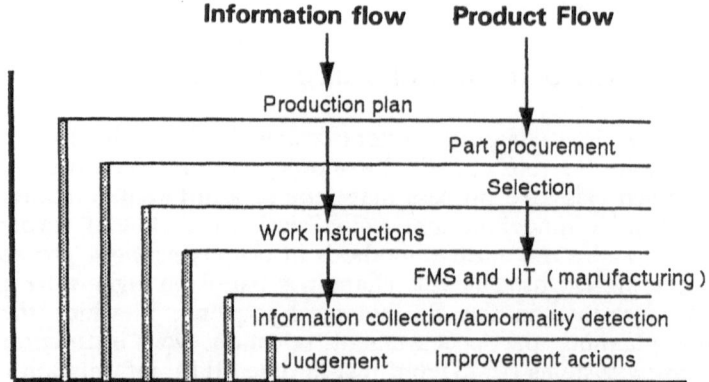

Fig. 4.2. Work processes of skilled workers in production activities.

4.1.2   Sequence of Products and Information

Figure 4.3 illustrates a typical information system for manufacturing, and shows the sequence of products and information. Sales offices and agents, located all over the country, receive enquiries from customers. If the customers are satisfied with the products and the prices, they place orders. Their orders are checked against stock. If some products are not in stock, the order information is sent to manufacturing sites. At the same time, if the customer's requests involve the development of new products or the modification of current products, the order information is transferred to the computer-aided design (CAD) department. The design department often participates in negotiations with the customer from the early stages, including enquiries. This process is the sales and distribution system. In the next step, the material requirement planning (MRP) system step, materials are procured and production plans are made according to order information received from the product sales and distribution system. Items of order information are rearranged according to priority in order to maximize efficiency in the production field. Job instructions are then given to the field at the appropriate time. In the manufacturing field, products are manufactured by FMS and JIT according to job instructions received from the MRP system. After inspection, the completed products and related information are returned through an MRP system to the product sales and distribution system, eventually reaching the customer.

4.1.3   MRP System and Work Instruction System

There is a close relationship between the MRP system and the work instruction system. The jobs of skilled workers in manufacturing can be classified into two groups: indirect manufacturing jobs and direct manufacturing jobs. Indirect manufacturing jobs involve intellectual work such as planning, whereas direct manufacturing jobs involve actual manufacturing processes, where workers use their hands and operate machines. FMS has made it possible for unskilled workers to handle direct manufacturing jobs. Now, with advances in automation, productivity improvement depends largely on enabling unskilled workers to participate in indirect manufacturing jobs. In Sect. 4.2 examples of indirect and direct manufacturing job systems will be introduced.

MRP systems have been used by many companies as general-purpose computer systems with a wide range of applications. Generally, MRP systems have a variety of functions, including production planning, part procurement, and selection, which had previously been controlled through batch processing. However, the speed of planning has recently been slow relative to demand in the production field. There are now two major trends in the manufacturing industry: one is the production of a large range of items in small quantities, and the other is the production of a small range of items in variable quantities. Both trends have made production lots smaller, changing the frequency of production planning from weekly to daily, and from daily to hourly (real-time scheduling). In short, production planning must now be carried out in real time. To conduct manufacturing efficiently in this environment, it is necessary to consider many fluctuating

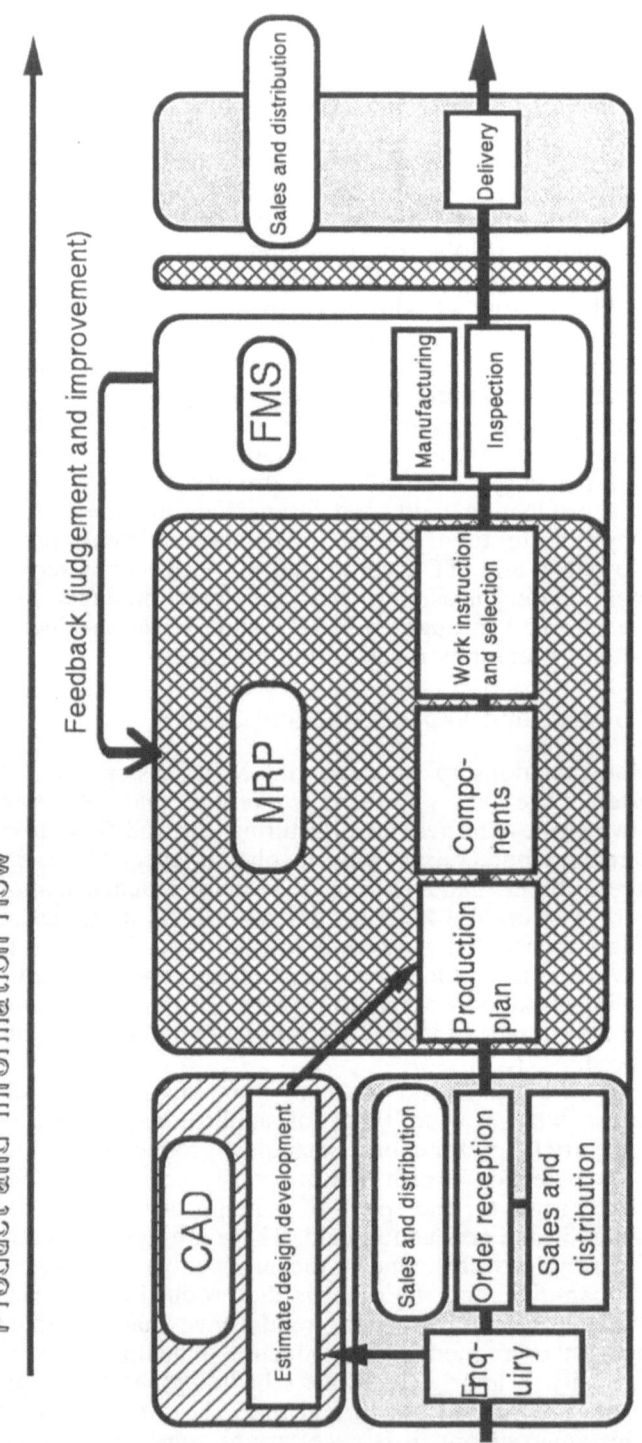

Fig. 4.3. Relationship between products and information.

elements, including production line capacity, operation rate, and progress in the manufacturing field. Thus, more frequently than ever before, production plans are modified in the field. Plans must be modified in real time, and it is vital to consider conditions in the production field. For example, if a robot fails, the production plan must be changed on site, according to the scale of the failure. This requires experience and knowledge, and must be left to intelligent experts. Furthermore, a great deal of work is involved in making plans in real time. This task is difficult but necessary for improving productivity.

Figure 4.4 describes the need for real-time plan modifications. In this case, the system is divided into three parts. The mutual relationship between these parts is also shown. From the top, they are an MRP system, the production instruction system, and the manufacturing information collection system. As previously described, it has become difficult for an MRP system on its own to respond to frequent plan modifications in the production field. This would not be a problem if MRP systems could make plans in real time. In reality, however, hardware restrictions inhibit real-time planning features from being implemented. Thus, it is necessary to establish "a mechanism for efficient job instructions". What should be done to minimize plan modifications? The answer is to make plans as late as the material procurement schedule allows. If, for example, plans are completed immedi-

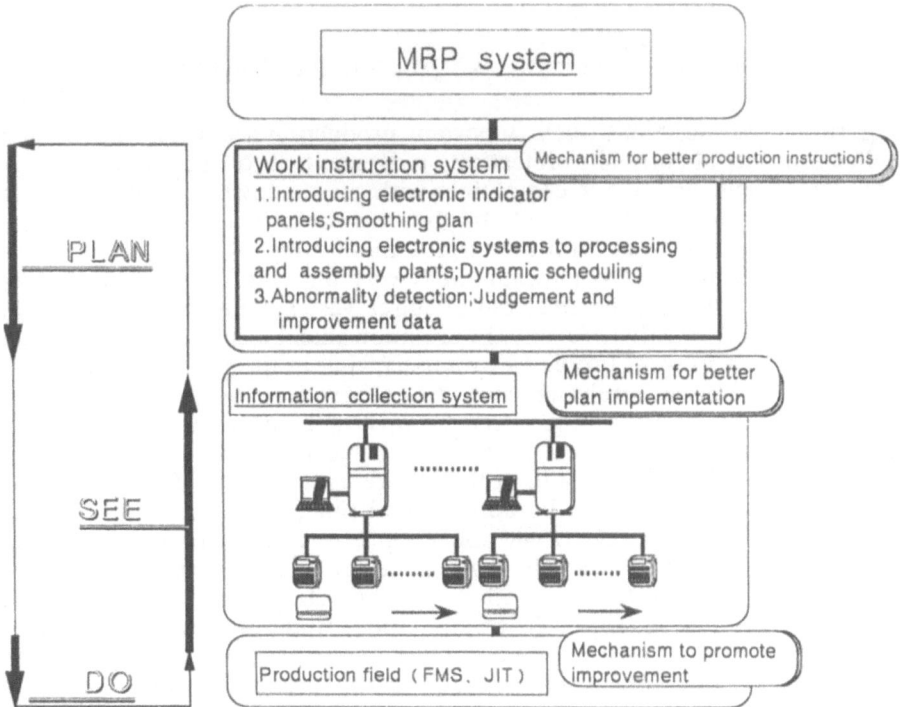

Fig. 4.4. Relationship between production management, work instructions and information collection.

ately before production, no modifications are expected. Plan modifications are not created in the production field; they are created in the production planning department. It is, therefore, desirable to combine the MRP system with the manufacturing information collection system. On one hand, the MRP system creates macro plans such as weekly or daily production plans. On the other, the manufacturing information collection system collects production information in real time. The manufacturing information collection system receives a macro production plan from an MRP system and gives information on the current status of the production field, including progress, material processing, and robot failures, to an MRP system. Using that information, job instruction plans can be created in the shortest possible time so that all parts of the production line are utilized. Such a system is called a job instruction system. This is a computer-based dynamic scheduling system. Likewise, minimizing the time required between planning and job instructions minimizes production plan modifications. This system eliminates the need for skilled workers with complex knowledge and experience. The replacement of skilled workers by expert systems is desirable. Computer systems based on AI technologies will provide these expert systems.

Future progress in technology and the development of faster, cheaper computers will enable MRP systems to make plans in real time. At present, however, it is effective to divide the real-time planning process into a production planning system and a job instruction system (dynamic scheduling system).

4.1.4   Information Collection and Abnormality Detection

Figure 4.4 also describes other important production activities: information collection and abnormality detection, judgement, and improvement. The abilities of humans are not compatible with these activities. Manufacturing information can be collected only by thorough observation of a wide range of production processes or by constant monitoring of the operations of advanced machines. Such information is very important for understanding the essentials of the problems, but requires vast amount of energy, time, and experience to gather. Ideally, information shall be collected and utilized to minimize or prevent problems from occurring in the production field. In reality, information rarely reaches the production line supervisor (line leader) who has the techniques and know-how to prevent problems from occurring. In addition, most production line supervisors are too busy to collect information. As a result, problems are handled only after they occur. Information collection must be done in real time, and this is not practical for human beings. As electronic tools, such as computers and networks, become less expensive, systems using these tools can be constructed for a relatively small investment. Skilled workers start solving problems by collecting information. Similarly, creating jobs for unskilled workers requires a computer system for real-time information collection [2,4]. It is also important to establish a method to reveal problems when designing production mechanisms. If methods for revealing problems and collecting information are established, improvements will become obvious. Currently, however, most functions performed by these mechanisms are left to skilled workers, except in the MRP process. Thus, the most important topic in this chapter is how to introduce a computer system into this field.

4.1.5 Judgement and Improvement

Judgements can be made by correlating collected information with knowledge and experience. Based on this judgement, conclusions can be made or execution plans created. In this area, numerical expressions and simple theories rarely count. The decisions of experts are what really count. Activities in this field tend to rely on human qualities, especially past experience and knowledge. This area is very important for improving productivity by eliminating bottlenecks and slackness in the production process. Efforts spent in this area may well determine productivity. This is the most important factor in computerization, and thus requires advanced technologies. Artificial intelligence systems such as expert systems and fuzzy logic systems will be fully utilized. This can be seen as the converting of the jobs of intelligent and skilled workers into a computer system.

When the information collection and abnormality detection systems described here are combined with judgement and improvement activities and the participation of production staff, improvement will be rapid and production will be optimized. This also allows unskilled workers to perform skilled jobs, thus enabling spontaneous, stand-alone production activities.

4.1.6 Section Summary

The jobs of skilled workers include more than just the manufacturing techniques of expert manufacturing staff that can be replaced by an automated FMS. Many jobs are carried out by skilled workers who are not directly involved in manufacturing. Such jobs are called indirect manufacturing jobs. They include production planning, work instructions, and manufacturing information collection. Recently, replacing these skilled jobs with computer systems has become very important. This section has discussed methods of making the planning process dynamic. Integrating improvement processes with information systems in which past information has been stored helps eliminate bottlenecks and slackness from the production process. This enables higher productivity and superior quality. Such systems, in which improvement activities are integrated with information systems (such as job instructions, information collection, and abnormality detection systems), are called "production improvement systems". Production improvement systems form a field of CIM, and are significant methods of future productivity improvement. Furthermore, these systems are designed with the user in mind and are worker-friendly. From the general views mentioned, we can expect to gain basic knowledge on unskilled worker-oriented manufacturing; however, case studies of such manufacturing still enable us to increase our knowledge. Thus, two typical case studies, conducted by Omron Co., Ltd. of Japan, are described in the following section.

4.2 Case Studies on Unskilled Worker-Oriented Manufacturing

4.2.1 ALFA Systems in Ayabe Plant

Ayabe Plant was established to manufacture sensors. A fast local area network (LAN) links the entire plant and handles the production of numerous items in small lots. The LAN is used for job instructions and information collection. To reduce skilled workers' jobs in both the direct and indirect manufacturing jobs, the ALFA automation and manufacturing information management system is used throughout the plant. [1]

Concept of the ALFA System

The ALFA system fulfils the following manufacturing and operation requirements.

1. *A Tool for Producing Numerous Items in Small Lots.* It is generally acknowledged that the demand for sensors is growing, and various types of sensors have been developed. In the field of sensors, customers expect to be able to buy as many sensors of their choice as they want. To respond to such market demand, manufacturers must manufacture products in small lots upon receipt of orders, rather than waiting for orders to accumulate before beginning manufacture. Furthermore, the manufacturing system must have a short lead time to respond to changes in user demands.

2. *Ease of Use for Everybody.* Ayabe Plant started operation in April 1986, when the production staff were all new employees, in other words, unskilled workers. There was little time for training, so the development of the ALFA system took ease of use into account. The aim was to develop a simple, clear system that anybody could use.

3. *Human-Oriented System.* In the ALFA system, the computer operates in response to the actions of the workers. This allows the workers to retrieve necessary information from the computer, and to carry out production and improvement activities by operating the computer as required. In this system, the computer does not use people; people use the computer. In this way, the production system does not depend on the know-how of the system developer; additional developments can be made as required through the know-how of the users. For this purpose, parameters are set via input and output from factory automation (FA) cards and cathode-ray tube (CRT) displays.

4. *Real-Time Monitoring of Production Field Statuses.* Various information is necessary for plant management, including production progress, quality, yield, and machine maintenance. Such information is meaningless if it is not available for access in real time when needed. Collection of data is often carried out for its own sake. In addition, data are usually collected and reported by people in the manufacturing field, who may distort the

values. Thus, we specified minimal data items to understand the real statuses of the production field, and designed the ALFA system to sample these items fully and automatically.

5. *Not a Technical Model, but a System that is Economically Feasible.*   We formed our concept by collecting opinions from many people, inside and outside the company, and by asking questions about their FA concepts and systems. However, we could not find an FA model that could be applied to our assembly line, so we had to develop an original system. For our purposes, we recognized that the original FA system should prioritize economic feasibility, rather than pursuing a technical model. Based on this, we developed our budget.

Configuration of the ALFA system

The major components of the ALFA system are factory and line computers (i.e., production line computers), workstations, LAN controllers, data collectors (D/C), and a LAN. See Fig. 4.5 for a description of the configuration of the ALFA system. The major components are as follows [5,6].

1. *Network.* The factory computer and workstations are Omron-made UNIX machines connected to an Ethernet. The line computer and data collectors are connected to another Ethernet. This stabilizes the data throughput by separating the office automation (OA) network from the FA network. For example, in general, small-volume data are repeatedly exchanged in short cycles in FA, while large-volume data are irregularly exchanged in OA.

2. *Data collectors.*   During developing, there were no appropriate terminals (peripherals) available in the market. We decided to use FA personal computers as terminals in the production process. These terminals are connected to the FA Ethernet via the LAN controllers.
   A data collector has built-in firmware that works as a man–machine interface and a machine–machine interface. Through the firmware, bidirectional manufacturing data are exchanged between the data collector and the line computer. NC machine, manual assembly, storage, and other attributes can be assigned.

Flow of Production and Information

The daily production plan is downloaded from the mainframe MRP system to the factory computer on the night before production. The plan is then classified for each job area code, and further downloaded to the line computer on each floor. At the beginning of the day, the production plan is linked with the FA card for each order, and entered into the lines. In the ALFA system, information is integrated with products by sending FA cards with trunk parts (printed-circuit boards (PCBs)) through the production process from beginning to end. FA cards (see Fig. 4.6) are the same size

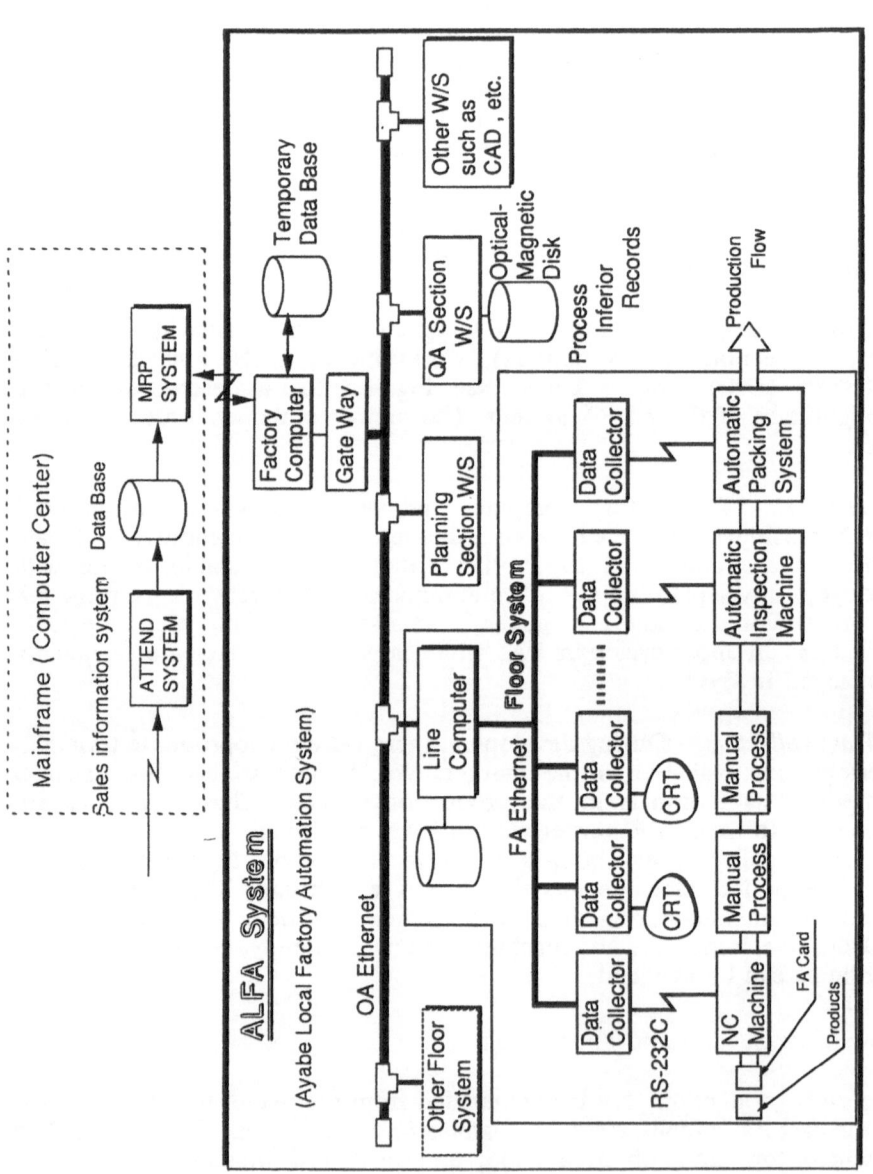

Fig. 4.5. Configuration of the ALFA system.

as PCBs (2 in square) and are made of stainless steel so that they can withstand heating (such as from solder reflowing).

FA cards have numerical codes punched on their surfaces that are unique to each plant and are fitted with MRP order numbers at the start of production. At the beginning of production, the reader in the data collector reads the FA card number. Depending on that number, the line computer automatically downloads the necessary production information to the data collector and the production facilities (see Fig. 4.7, data download sequence). At the end of each process, the results are uploaded interactively to the line computer.

Unnecessary in-process stock can be prevented by limiting the number of FA cards issued. In recent fully automated assembly systems, identification (ID) systems are used in combination with FA cards. This enables perfect single-item production flow.

Information Handled in the ALFA System

In the ALFA system, the information flow is as follows.

1. *Information Downloaded from the Line Computer to the Data Collectors.*
Through data collectors, a line computer sends the production instructions required for the staff and facilities in the production process. The major content of the instructions is as shown in Table 4.1.

2. *Information Uploaded from the Data Collectors to the Line Computer.*
At the end of production, a record of interior processes (PIR) is uploaded to the line computer and stored with MRP header information. The information shown in Tables 4.2–4.4 in the record is stored in the following sequence: production line, year, month, day, and order. The information is used as backup data for the quality assurance (QA) section and production improvement. The fact that information related to production, facility operation, and quality can be collected in real time and recorded automatically is in line with the ISO9000 concept.

Actual System Examples

In Ayabe Plant, there are two different production lines for assembling switches in the ALFA system. The first line is used to assemble contactless

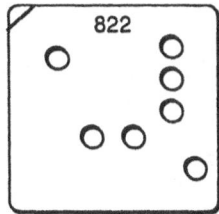

Fig. 4.6. FA card.

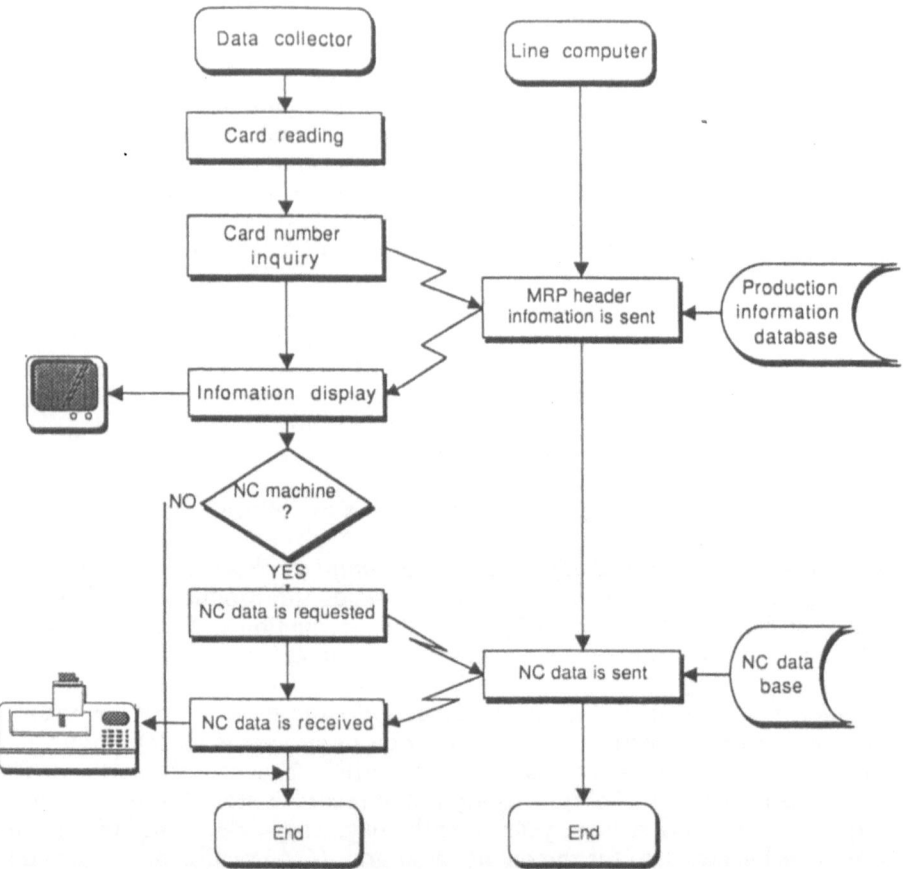

Fig. 4.7. Data download sequence.

switches while the second is used to assemble photoelectric switches. Their layouts are described briefly in the following.

1. *Mixed-Flow Assembly Line for Contactless Switches.* Figure 4.8 shows the configuration of a contactless switch. User demands for contactless switches keep changing, thus there are currently over 10 000 product types. This is a typical example of the production of numerous items in small lots. Figure 4.9 shows an actual contactless switch production line. This line produces 650 major models and 1500 different specifications of contactless switches alone. Production proceeds according to the information recorded on the FA cards issued for PCBs, which are fundamental components. Note that no human intervention is required for these automated processes, because production conditions are changed automatically. This mixed-flow assembly line for contactless switches has a very short lead time and a small amount of in-process stock, as shown in Table 4.5, enabling production to proceed according to schedule. In both indirect and direct manufacturing jobs, skilled workers' jobs have been greatly reduced. (*Note*: Ayabe Plant

Table 4.1. Production instruction information

Name	Description	Example usage
Order number	Order number for MRP management	Orders covering more than 40 items are split so that each order covers only up to 40 items
Model code (Product number)	The highest code in the MRP part configuration tree	Automatic stage switching information for automatic assemblers and automatic inspector
Production model format name	Model information for workers	Product format information
Manufacturing specification	Manufacturing information for workers	Information such as power supply specifications and cable lengths
Planned production quantity	Instructed production quantity	The average yield rate is added for each format The minimum quantity for instructions is one item
Deadline	Planned production completion date	
Comment	Design modification information	This information is not defined
Facility data	NC data	Registered in the line computer for each model and facility

Table 4.2. Production information

Production quantity	Actual quantity produced
Production time stamp	Time of the production and manufacturing history
Lot numbers of significant parts	Lot numbers of the parts used in the product
Facility name and number	Name and number of the facility that manufactured the product

Table 4.3. Facility operation information

Operation time	Time for which the facility operated
Stop time for each reason for stoppage	Time during which the facility stopped for each reason
Facility operation rate	Facility operation rate

Table 4.4. Quality information

Dropout quantity	Number of defective products
Number of defective products for each reason for defect	Number of defective products for each reason for defect

Table 4.5. Effects of the contactless switch production line

Production lead time	Shortened to 30%
In-process stock	Reduced to 50%
Skilled workers in direct jobs	Reduced to 50%
Skilled workers in indirect jobs	Reduced to 30%
Defective products	Reduced to 50%

is a new plant. The decrease in the number of workers is calculated by comparison with the number assumed to be needed for conventional production methods. Thus, no workers have lost jobs in this plant.)

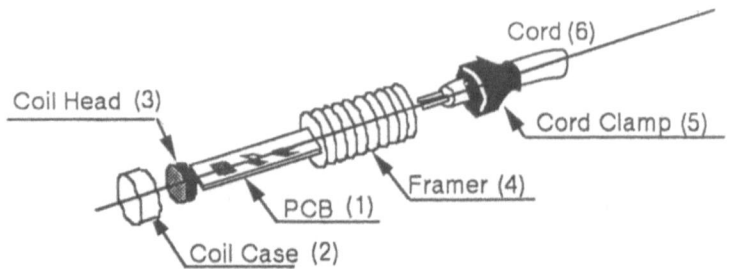

Fig. 4.8. Contactless switch configuration.

2. *Fully Automatic Assembly Line for Photoelectric Switches.* This line consists of 17 standardized assembly cells (see Fig. 4.10), linked in series (see Fig. 4.11). Unlike contactless switches, some photoelectric switch models consist of a set of a projector and a receiver. Thus, in some cases, two different products must be manufactured to make one complete set. For this reason, an ID system is introduced in this line in addition to the FA card system. Order information is managed according to the FA cards, and information on individual products is managed via the IDs. Thanks to the combination of these two systems, it is now possible to manufacture products with complex specifications both continuously and independently. In this production line, an ID carrier is attached to every assembly pallet on which product components are assembled. The ID carriers contain manufacturing information on individual products, and the robots read that information and assemble different products for each assembly pallet. This fully automatic assembly line for photoelectric switches has almost completely eliminated skilled workers' jobs and complex job instructions. Most jobs are now easy even for unskilled workers, and the number of defective products has decreased.

Outline of Mishima Plant

Mishima Plant produces PLCs (programmable logic controllers) and FA computers which are the key components of total FA (flexible automation or factory automation). At present about 2000 types of PLC are produced, and about 70% of the standard products listed in our catalogue are ready for immediate delivery. In response to this demand, the production line uses mixed-line production, a system ideal for the production of a medium number of items in medium-sized lots. The two computer systems designed for unskilled workers in this plant are described in the following.

Automation of Selection Instructions in the Component Warehouse. In Mishima Plant, selection, which involves collecting necessary components from the warehouse, can be left to unskilled workers, although such a job affects the product quality and yield rate to a large extent. In the production of electronic components such as PLCs, labour concentration is the key. To ensure a high quality of assembly, Mishima Plant employs kit distribution production, which utilizes the components warehouse and procures the necessary quantity of components in kits. In the warehouse, components

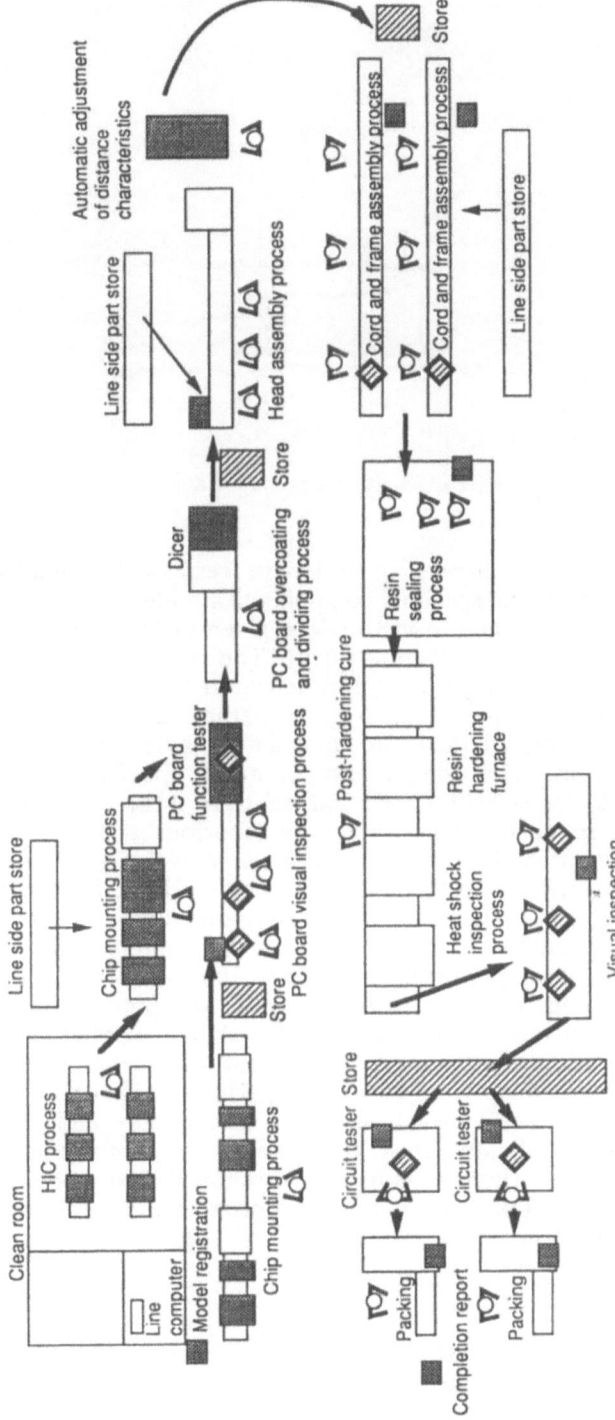

Fig. 4.9. Contactless switch production line.

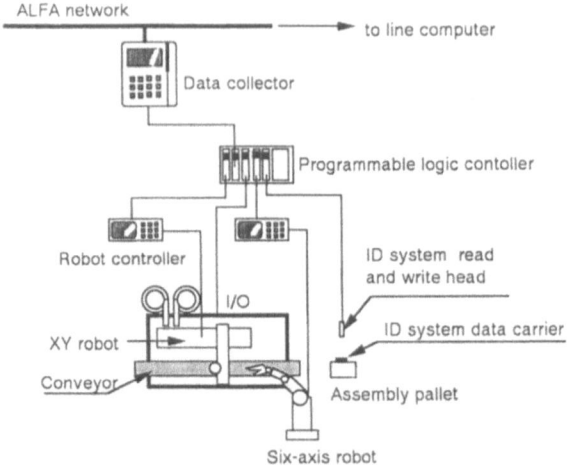

Fig. 4.10. Assembly cell configuration.

are delivered as follows: in order to respond to multi-item small-lot production lines, the specified components are identified out of about 12 000 components, according to instructions given in the selection lists which are output daily (200 to 300 orders/day). The components are then classified for each order and work area and delivered to the production line.

Previously, this was done by the method described in Fig. 4.12. The components were divided and assigned to a few workers. Each worker found their assigned components from the specified component shelves according to the selection lists, and transferred the components to the next worker in sequence. This previously laborious task now consists of the elements described in Fig. 4.13, which classifies the selection jobs into those for skilled workers and those for unskilled workers. From the top, jobs for skilled workers include checking the component list, finding the shelf where target components are stored, finding the components, and putting the components in component boxes. Jobs for unskilled workers include counting the components, packing the components, and putting the packages in the containers. In this system, the jobs for skilled workers are now done by computer. Comprehensive experience and knowledge is required for each element. To develop from being an unskilled worker to a skilled one, the worker has to be knowledgeable on components and receive extensive training on installation. Thus, the system aims at maintaining productivity with unskilled workers by computerizing the skilled workers' jobs, namely the jobs involving the selection of appropriate components in the warehouse, as described in Fig. 4.13.

Outline of the System

According to the production plan, material requirements are calculated by an MRP system, and job selection diskettes are created. The data on the diskettes are classified by the FA computer for each rotary stocker and

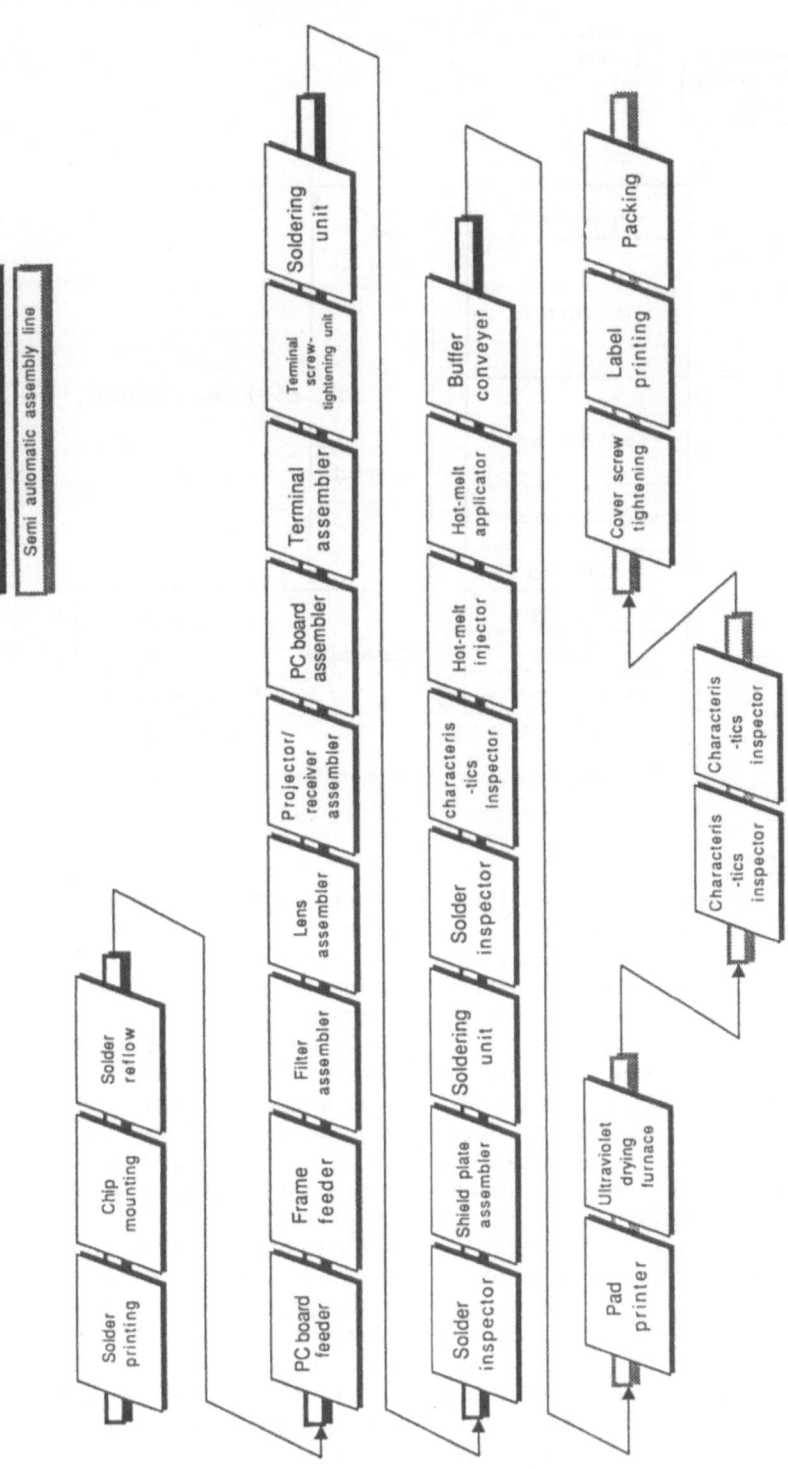

Fig. 4.11. Configuration of fully automatic assembly line for photoelectric switches.

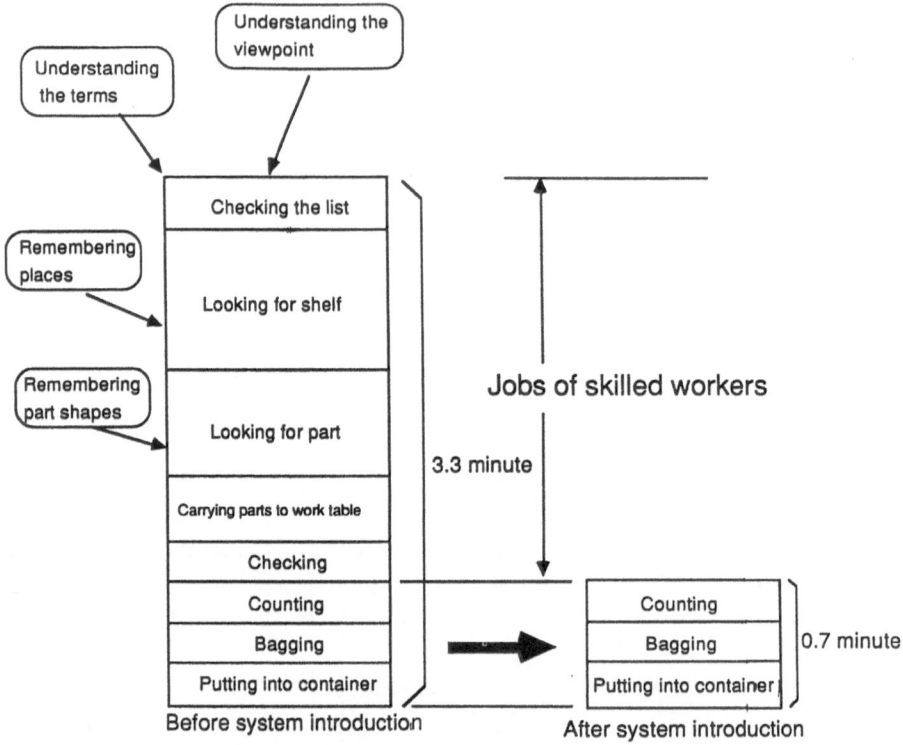

Fig. 4.12. Mechanism before introduction of the system.

order number. The FA computer then transfers the data to the SBC (single-board computer). The SBC converts the data into a control data format that matches the PLC, and transfers the data through optical fibres to the 16 PLCs connected. This creates a computer system that gives complex selection instructions quickly and accurately, rather than using skilled workers. First, the worker presses the start button. Then the rotary stocker shelves rotate, positioning the appropriate shelf in front of the worker. The shelves automatically stop rotating and the indicator light goes on to indicate the block that contains the specified components. At the same time, the display unit beside the worker displays the order numbers, the number of items in stock, and the number of remaining selection operations, and issues a delivery label using the label printer. The worker removes the necessary number of components according to the delivery label, and presses the delivery completion button. Before the next components are selected, the worker packs the components in a bag, appends the delivery label, and puts the bag in the container. If the display unit indicates that no selection operations remain, the selection operations of the day are complete. Figure 4.14 shows the system configuration, and Fig. 4.15 shows the selection system. In addition, Table 4.6 summarizes the significant effects of the instruction system for automated selection.

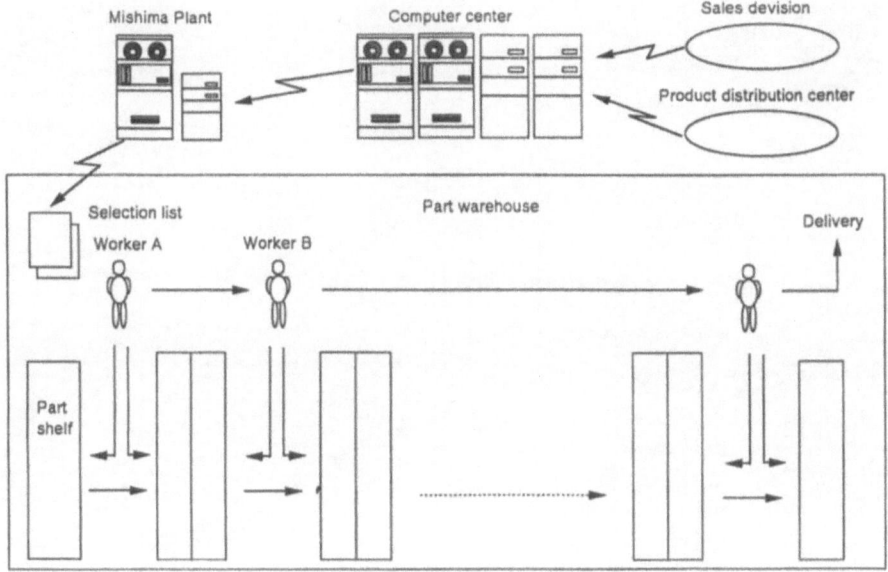

Fig. 4.13. Jobs of delivery workers.

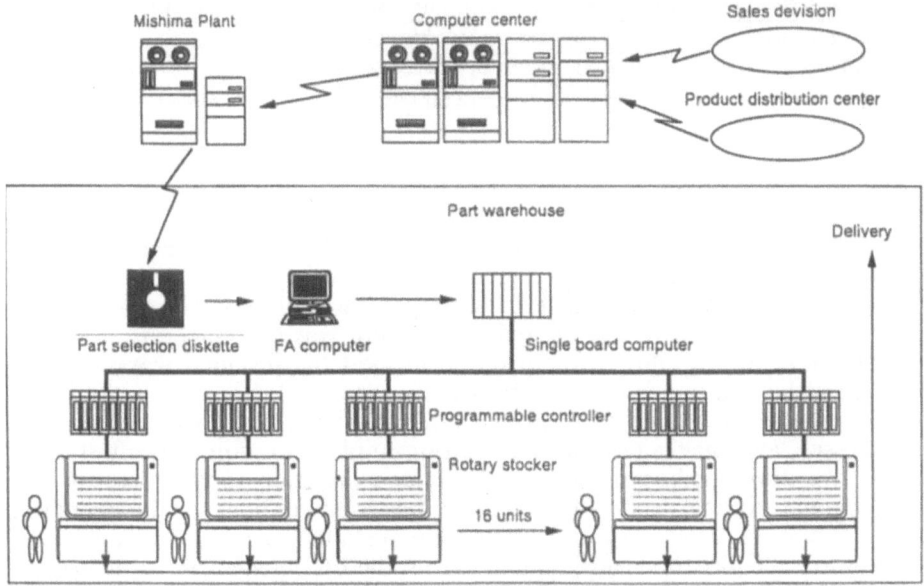

Fig. 4.14. System configuration.

Fig. 4.15. Selection system.

Table 4.6. Effects of automated selection instruction system

Skill requirements of the job	Education is no longer necessary. Unskilled workers can do the job.
Job improvement	Delivery mistakes have been reduced to 1/20.
Job efficiency	The selecting speed is increased by five times.
Other effects	The walking distance for selecting has been shortened from over 10 km/day to 1 km/day. 40% of space has been saved.

Dynamic Scheduling System in the Comb-Shaped FMS Production Line for Mounting PCBs. The comb-shaped FMS production line is designed for assembling PCBs of different process sequences and cycle times in mixed-flow production. A dynamic scheduling system was introduced in the FMS production line to allow jobs to be performed by unskilled workers. The FMS production line is outlined below. In ordinary linear automated lines, producing several types of product causes great losses in balance between processes, and makes efficient automation implementation difficult. Therefore, we developed seven differently shaped insertion robots (six-axis XY orthogonal robots) and three manual insertion stations as independent assembly cells. These assembly cells were installed in a comb shape along the automatic monorail cart line that carries the PCB magazines. A magazine containing between one and 30 PCBs (as many as are specified in each order) is sent. The magazine has a bar code label attached to indicate the model. When the magazine reaches an assembly cell (mounting robot) station, the FA computer reads the bar code on the magazine to identify the necessary NC data. According to that data, the FA computer transmits the NC data on that model to the assembly cell (mounting robot) through the network. The assembly cell automatically arranges the assembly steps according to the NC data, and assembles the PCBs. With this type of production line, the manufacturing lead time is halved and productivity is improved by up to 400%. The number of jobs performed directly by skilled workers is reduced by 70%. Figure 4.16 illustrates the FMS production line process, and Fig. 4.17 shows the FMS production line.

This FMS production line is designed to handle the mixing of multiple items during production. It is, therefore, necessary to take into account the sequence or timing for entering the magazines, otherwise the flow of magazines in the line may stagnate, slowing down the operation of the total system. For example, if the same products are sent consecutively, some assembly cells are kept very busy, but other cells are left idle. This lowers the total operating rate of the system. To improve the operating rate of the system, it is necessary to understand the operating conditions of each assembly cell and decide, in real time, which product to handle next. This work requires experience and knowledge. Previously, the magazine entry sequence was optimized under the supervision of skilled workers, who used experience. We, therefore, developed a dynamic scheduling system based on expert system technology. Now, unskilled workers can handle the work thanks to the entry sequence instructions given by the dynamic scheduling system [3] (expert system for ordering introduction work: ORDERS). Figure 4.18 shows the configuration of this expert system, and Fig. 4.19 shows its appearance.

An expert system (KBMS/PC™, NTT Corp.) installed on Omron's FA computer (FC985™) consists of a knowledge base and inference mechanism. This personal computer-based system is very economical, and collects real-time information on production line operations, which is essential for inference, through the network. Data required for scheduling are automatically collected and input, reducing the burden on workers. The installed expert system has a fast inference algorithm, and completes inference while orders are being switched. This feature contributes to real-time scheduling. The system has proved to be as capable as a skilled worker with great experience and knowledge.

To implement a scheduling system, additional information must be obtained in real time. There are two items of such information: in-process order information and line entry queue order information. Both must be understood by the computer. According to these two items of information, the KBMS expert system, rather than skilled workers, quickly creates complex schedules as follows. First, production line order information is collected in the factory computer (FC983™) by the network system installed on the production line (SYSNET™, top left of Fig. 4.18). The line entry queue information is collected in the magazine management console (FC986™, bottom left of Fig. 4.18). These two items of real-time information are sent on-line to the factory computer (FC985™) incorporating the expert system. The factory computer (FC985™) monitors the conditions of the in-process items in the production line, and determines which order should

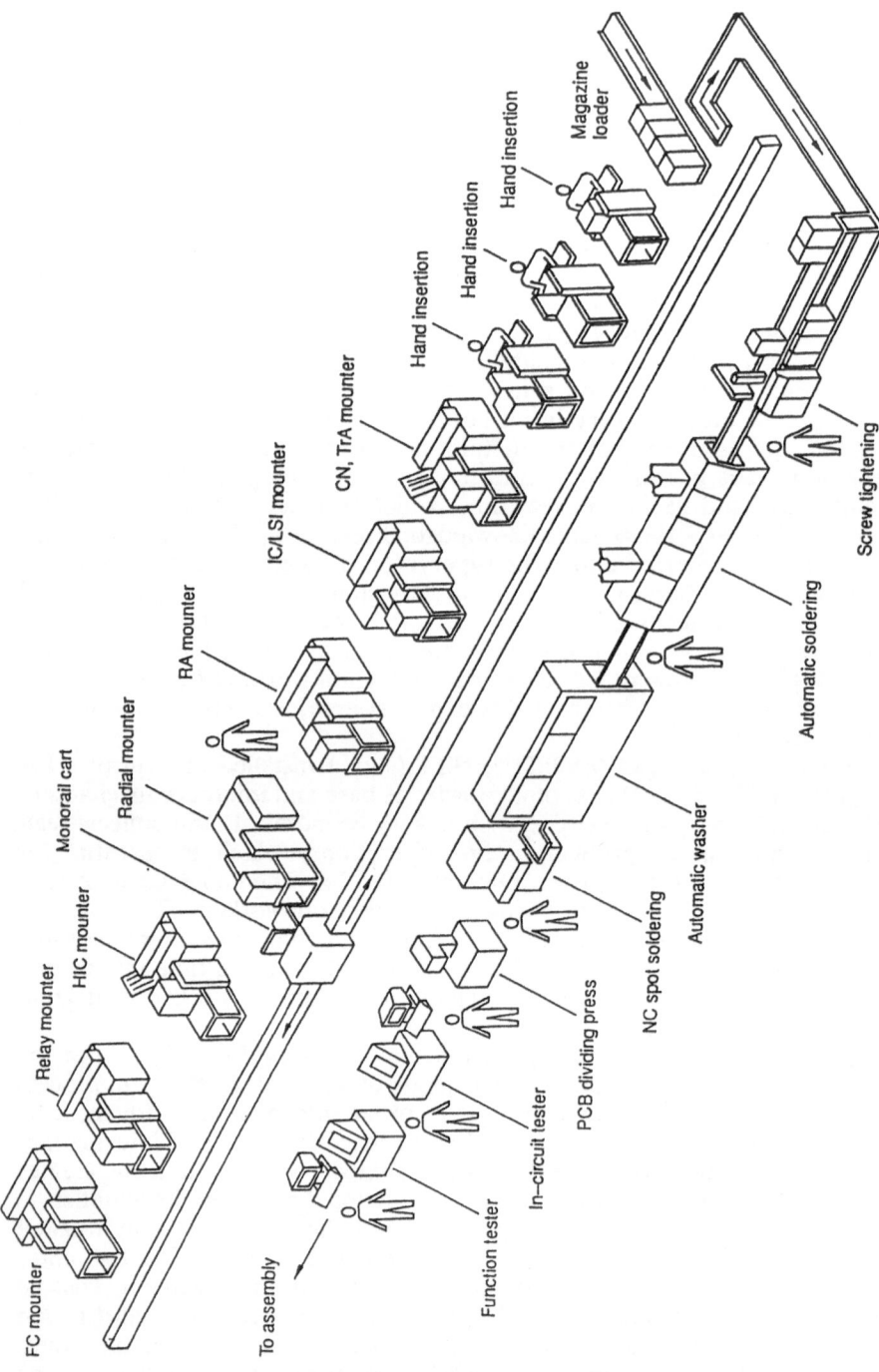

Fig. 4.16. FMS production line.

Fig. 4.17. FMS production line.

be entered into the line next to maximize the production efficiency. The answer is obtained almost instantly and reported to the worker. The effect of introducing the dynamic scheduling system has been that jobs that used to require two years of working experience can now be handled by workers having only one day of experience.

4.3 Summary

The decline in the working population is a serious problem in every company. This tendency is very noticeable in the manufacturing industry. Companies must recruit good staff in order to survive. Simplifying complex work and reducing work that requires experience gained over long periods are significant solutions to this issue. They are also a step towards introducing a comfortable working environment in the production field, allowing workers to enjoy their work. This contributes to the social value of the company. In Sect. 4.1, the jobs of skilled workers were classified according to the production processes involved, and the contents of each job were clarified. This section also explained the production improvement system, a concept that integrates information systems and JIT. Section 4.2 described examples of skill-requiring jobs that could be done by unskilled workers through the use of computer systems and machines, and the effects of the introduction of such systems.

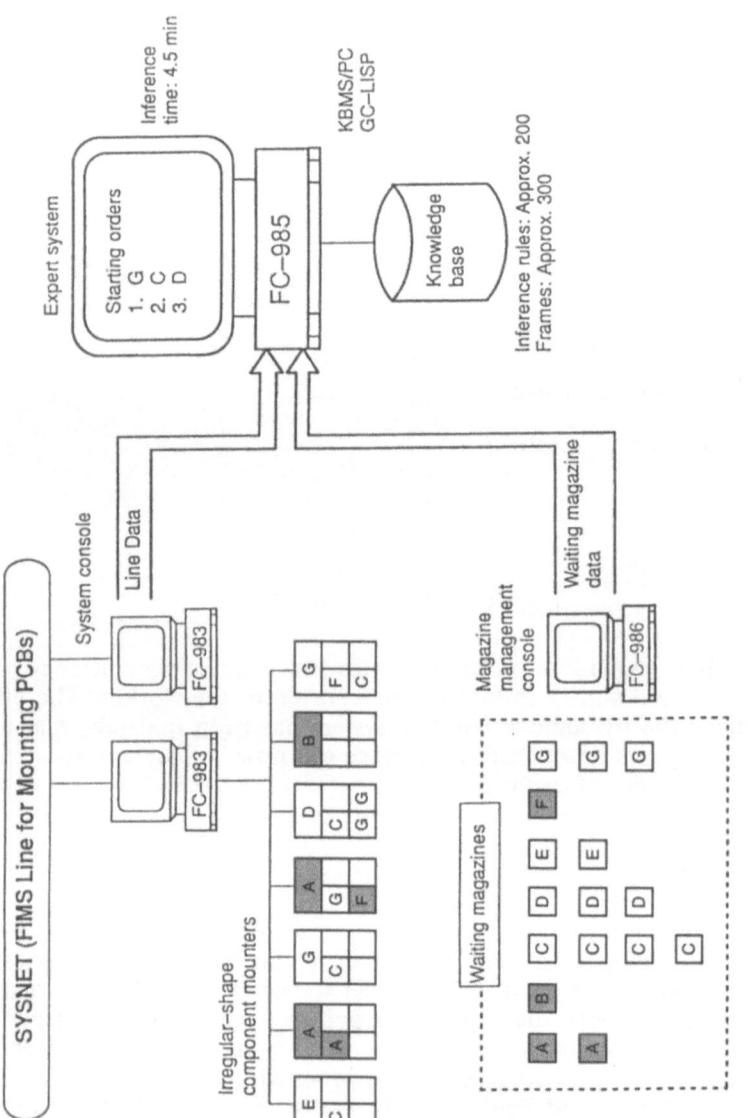

Fig. 4.18. Configuration of ORDERS.

Fig. 4.19. ORDERS.

Recent developments in technology have been very impressive. The methods described in this section will become even more sophisticated with further advances in computers and networks, and reduced costs in systems using these technologies. Growth in these activities may even determine the growth of companies themselves. We hope that the contents of this chapter are meaningful to engineers engaged in these activities.

Three people were involved in the writing of this chapter: Yoshio Mabuchi wrote Sect. 4.1, Kazuhiko Oishi and Koji Ono wrote Sect. 4.2. Jiro Iimura is responsible for the contents of this chapter.

Further Reading

1. Mabuchi Y. Ayabe plant automated with a middle-speed LAN. Comput Network LAN 1986; 37: 70–75
2. Mabuch Y. Example of Small FA-LAN. Appl. Biotechnol Chem Mech Electron Soc 1986; 2: 13–17
3. Mabuchi Y, Yoshii M. Dynamic scheduling expert system for FMS assembly line. In: Proceedings of the second intelligent FA symposium, July 1989; pp 137–140
4. Mabuchi Y, Ohishi K. Production monitor system. Trans Inst Elect Jpn 109-D (3): 1989; 150–152
5. Mabuchi Y. Local CIM system. In: Proceedings of the 34th annual conference of the Institute of System, Control and Information Engineers, May 16–18 1990, pp 521–522
6. Mabuchi Y. Local CIM system. In: 16th Annual Conference of IEEE Industrial Electronics Society (IECON'90), vol 1, Nov. 27–30 1990, pp 730–732

5 Flexible Computer-Integrated Manufacturing Structure of Global Network Type – Case Study in EC Countries

G. Spur and F. Zurlino

5.1 Introduction

The European economy is faced with important changes. After decades of slow movement within different countries of the European Community a single market was realized by the end of 1992. This means that Europe will turn into a territory without state frontiers, guaranteeing a free flow of people, goods, services and capital. An increasing opening of East European countries towards this development can be observed. It will cause considerable changes in the competition of business enterprises and industrial branches, as the sales potential as well as the number of competitors will increase to a large extent.

In the future, regional differences in customer requests will result in an even greater increase in product variety, so that many enterprises will be forced to revise their planning regarding production sites as well as production structures. Future production structures have to consider the expected changes and new site possibilities. The improvement of the factors of productivity, quality, flexibility and time can be achieved by a flexibly automated computer-integrated production approach (Fig. 5.1). Making use of site advantages as well as the presence of different markets will promote the development of overlapping production forms in factories and enterprises. Future-oriented structures require rapid transmission of information between sub-suppliers, customers and service enterprises by using technological aids over great distances. The infra-structure needed is already available and will be continuously built up over the years.

This chapter describes essential characteristics of computer-integrated production as it exists in many European industrial enterprises, and also describes the technological requirements necessary regarding communication facilities.

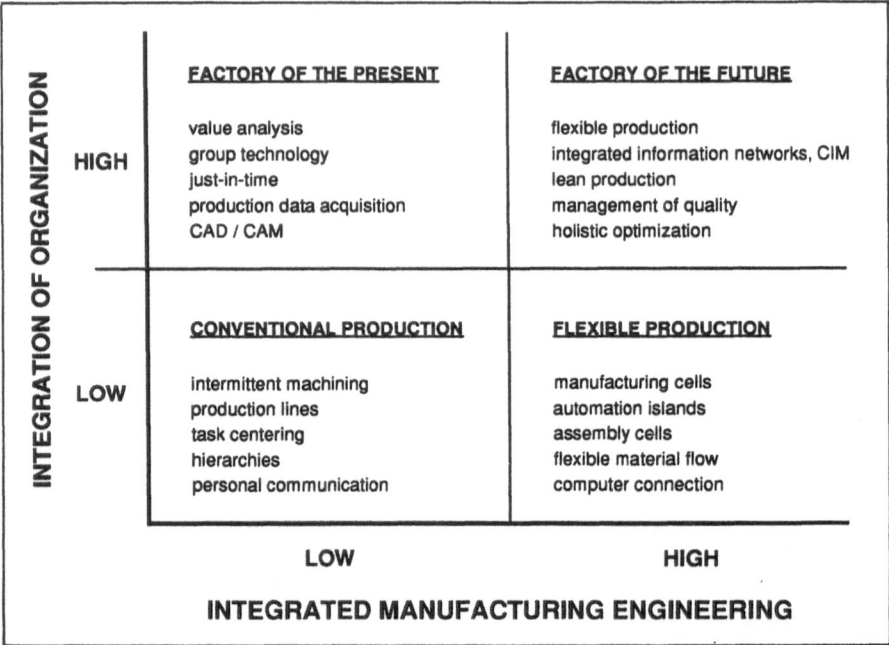

Fig. 5.1. Trends within the factory of the future.

5.2 The Computer-Integrated Factory

5.2.1 Factory Automation

The computer has been developed in order to carry out extensive calculations automatically. It has been realized as an instrument with the purpose of freeing man from the laborious repetition of analogous mental work. The motivation leading to mechanization and automation of material processes equals the motivation for the automation of information flow.

Automation can be conceived as a device to perform various types of repetitive work mechanically. With the development of information technology the means for effectively increasing production flexibility were provided.

The program-controlled computer is a generally applicable flexible control instrument. The connection of information technology and engineering is therefore the starting point for numerous further developments like numerical control (NC), computer numerical control (CNC), direct numerical control (DNC), industrial robots, flexible manufacturing and information networks. Observing the current trends, it is expected that these technologies will grow together to form one computer-integrated and flexibly automated unit, representing a further step towards the rational, humane and environmentally adapted production of vital products (Fig. 5.2).

Advantages of a computer-integrated manufacturing (CIM) concept are a higher production speed, flexibility, quality and reliability. The factory of the future will achieve a high level of industrial production technology,

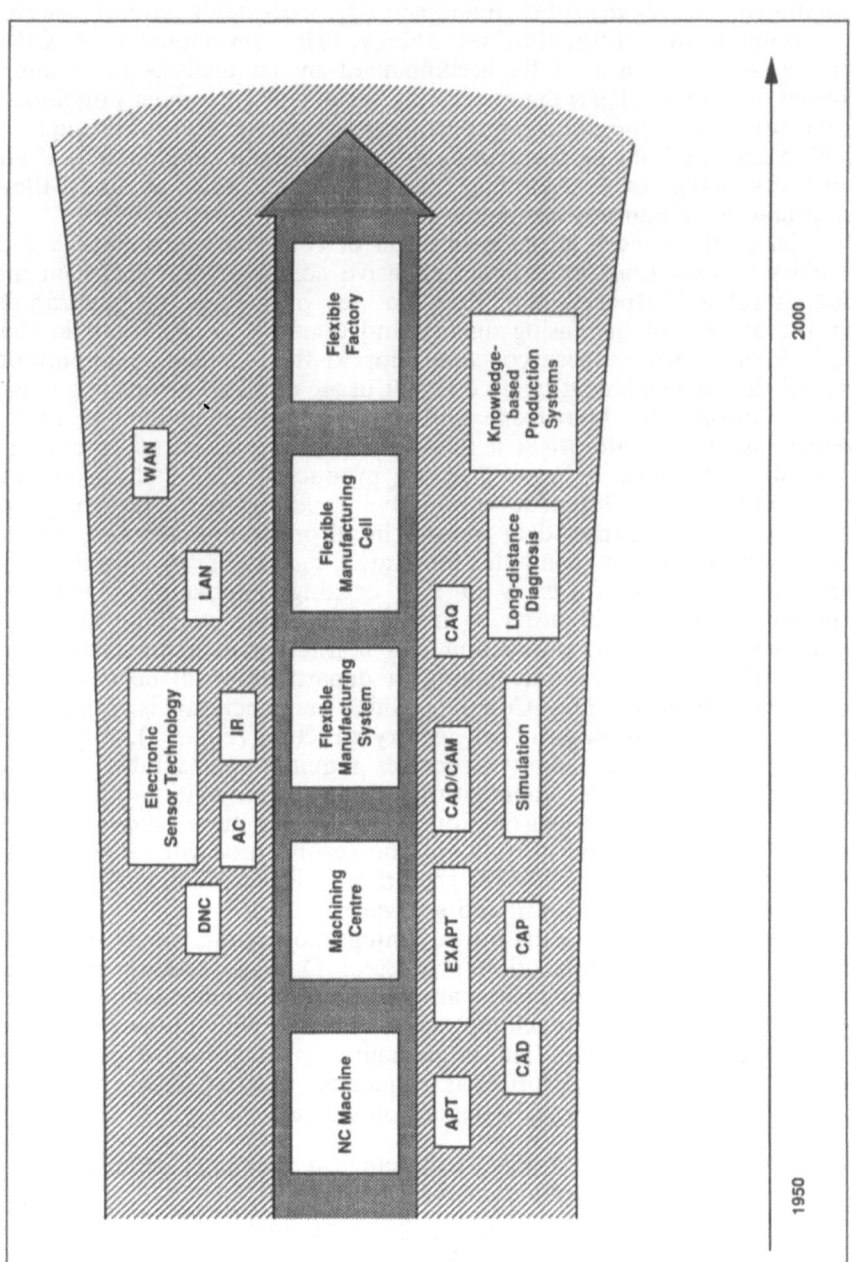

Fig. 5.2. Lines of development in flexible automation within 50 years.

which could not be realized with conventional technologies. Computer-integrated flexibly automated factory structures include a linking of all information processing machines, manufacturing processes, transport systems and computers by information technology. The development of CIM structures, however, has to be accompanied by an analysis of running conditions of traditional part manufacturing and assembly, company organization and production program. Efficient usage of the variety of information generated and used during the manufacturing process is only possible if all departments change over to binding conventions and processes and if they develop and use a uniform language.

The aim of the factory is the realization of concepts which contain the availability of new tools by a comprehensive computer application in all production-related departments. They do not only serve rationalization under the aspects of increasing quality and quantity. In addition, in the modern flexible factory these concepts support the increasingly important aspects of time management by an efficient usage of all capacities like data, facilities, information, knowledge and means of organization. In many European industrial enterprises a linking between the data of computer-aided design and those of parts program production is already taken for granted. Intensive development work is being performed in research and industry to enable a further data transfer in all production areas.

The transition to the computer-integrated factory is an evolutionary process. The CIM approach is to link existing automation islands by information technology in order to use the potentials of companies like development, acquisition, production and marketing to a maximum. An enterprise-wide computer-aided linking of departments will be developed out of various island solutions. Computer-aided manufacturing is an essential step towards decentralization of the factory structure (Fig. 5.3). A gradual development of existing factory structures requires considerable changes within three domains: a development of communication ability of hardware and software, changes of existing organization structures and a further development of engineering (Fig. 5.4). The second task will require great efforts from single companies, while the other tasks above all challenge the suppliers of machines, hardware and software.

Thus CIM is based on the linking and integration of the technical as well as the administrative information processes. Data generation and data processing equipment or machines are linked to the information flow in order to show possibly all working processes transparently in their available and non-redundant function. The implementation of advanced production structures in European enterprises is connected with technical, economic and also with human and social-oriented objectives, such as:

- Integration of tasks of design, technological planning, manufacturing control, production and quality assurance.
- Linking of machines, equipments, facilities, components and systems.
- Increasing the transparency of scheduling and organization within the whole enterprise.
- Decreasing the time for decisions and accounting.
- Fixing of definitive responsibilities for information content.

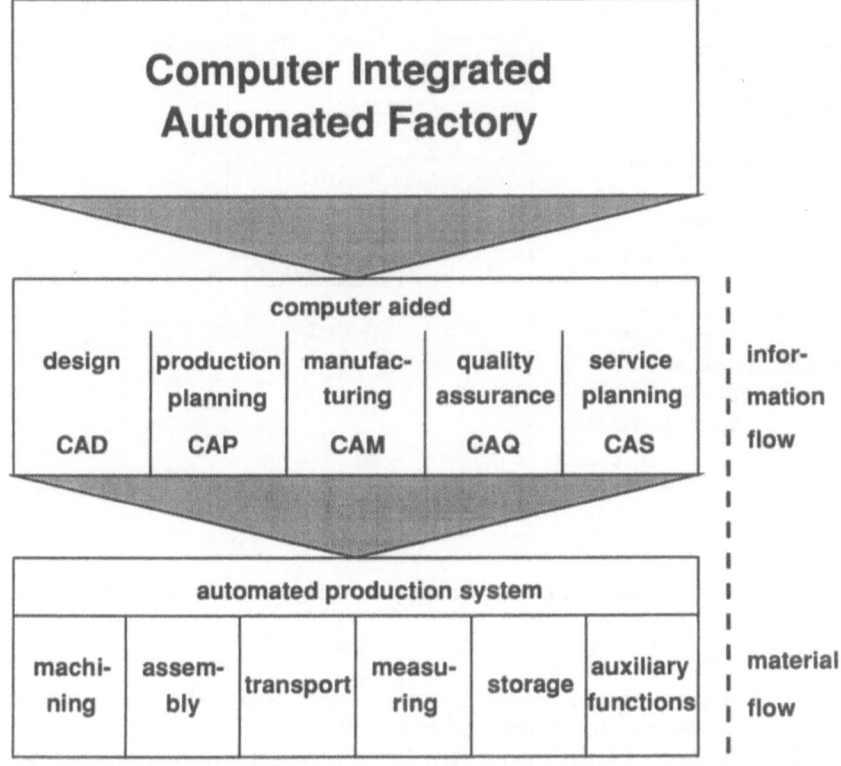

Fig. 5.3. Structure of the computer-integrated automated factory.

- Optimized application of computer programs through various possibilities of access and shared use.
- Realization of a comprehensive information flow throughout the company.
- Decrease of order stocks, throughput time, capital binding, and stocks of semimanufactured and manufactured goods.
- Higher capacity work load of capital intensive equipment by cutting down personnel in the second and third work shift, and adding week-end work.
- Increase of flexibility, productivity, faster reaction time.
- Improvement of product quality.
- Faster customization of products.
- Re-organization of work content in order to consider characteristics of human work.
- Improvement of methods and contents of staff qualifications and training.
- More complete usage of staff qualifications by optimized work content.
- Possibility of flexibilization of working time.

The computer-integrated factory organization enables pervasion of different domains by computer technology and a simultaneous concentration on information content. Changes occur within the existing technical and

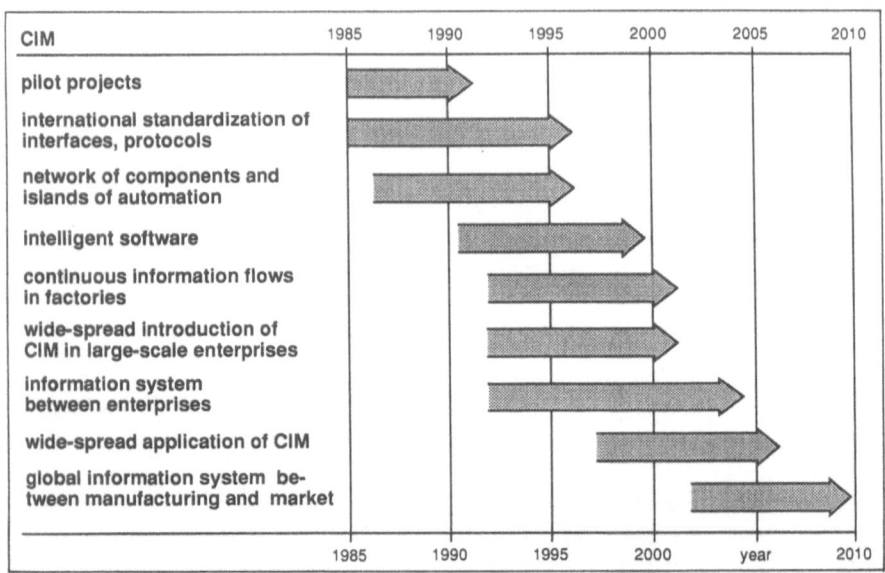

Fig. 5.4. Introduction periods for computer-integrated manufacturing.

administrative information flow regarding presentation and storage, but also regarding distribution, topicality and reliability.

The introduction of CIM causes step by step changes in the factory which offer defined interfaces for a widespread program structure. This requires digitalization of the whole information flow and a general availability of data, and if necessary, the establishment of commonly used data banks. Thus the creation of computer-supported and integrated production structures requires detailed preparation of current and stored data in order to achieve an orchestrated handling of tasks of the whole production process also in economical terms.

5.2.2   Flexible Manufacturing Systems

Parallel and in interaction with the technological development the requirements of the market have different effects on the aims of automation in separate areas of the European capital goods industry. In the field of unit and series production the technological and economical conditions are marked by an increasing product complexity with shorter product life cycles in parallel. Moreover, a larger variety of variants, shorter delivery dates and higher manufacturing quality are expected. As a result of this situation, different concepts of flexible automation were developed, due to the batch-sizes to be manufactured. These concepts, which differ in structure and application, consist of the following subsystems:

- Machining system
- Material flow system

- Information flow system
- Energy system

Flexible manufacturing cells are single machines with corresponding, peripheral facilities, which enable an autonomous operation for a certain period of time. The design of the flexible manufacturing centre is aimed at a machining centre, which can be adapted to very different user requirements, e.g. as a stand-alone machine or with the target of stepwise integration into a flexible manufacturing system.

Control systems of flexible manufacturing cells can be used as a CNC control with shop-floor-oriented programming functions as well as in connection with a host computer in DNC operation. Such a host computer does not transmit single data records to the stand-alone machine control, but entire programs. The transfer can be performed while the operation is running. The levels of performance of the control systems installed are still growing. According to the concept of distributed intelligence the aim is to integrate further functions from local CIM networks, e.g. continuous status control, functions of diagnostics, quality assurance and restarting functions after deadlocks.

Flexible manufacturing lines or transfer lines can be retooled for new manufacturing functions while they are out of operation. Each station is located behind the other with deeply shared manufacturing operations commonly realized. Flexible manufacturing systems emerge from connection of manufacturing cells and computer controlled machines, which are linked and connected with an automated material flow system. They allow a fully automated manufacturing process; moreover it is possible to control different operations on different objects within a defined spectrum of workpieces.

Flexible manufacturing systems cover the largest spectrum of application as they are suitable for single-piece production as well as for large scale production. These devices can therefore be characterized as a novel machining concept of flexible automation. Depending on the form of the workpieces to be produced they are distinguished in systems for prismatic and rotation-symmetric workpieces. In addition, flexible manufacturing systems have the following characteristics:

- The machining systems are linked by a central controlling system.
- The workpiece transport is performed in free intervals.
- Concerning the manufacturing method, manufacturing can take place on replaceable or different machining systems.
- Manufacturing sequences are variable.
- A flexible manufacturing system contains buffers for the adjustment of different operation times and for the storage of workpieces.
- The set-up of each machining centre can be performed free of interruption.

Flexible manufacturing systems differ particularly from flexible transfer lines or flexible manufacturing lines by the inner interlinking of ancillary machines implemented there. This manufacturing concept is suitable only for a limited spectrum of workpieces, for which a very short throughput time can be achieved, however. A set-up of one manufacturing unit causes a deadlock of the whole system.

The first flexible manufacturing system in Germany commenced operation in the early 1970s. In the beginning the number of installations increased quite slowly, but has accelerated considerably since the early 1980s. In the late 1980s approximately 400 flexible manufacturing systems for cubic workpieces were active in Europe, and trend rates show continuously rising numbers of installations.

Users of flexible manufacturing systems belong above all to the machine building industry, the automobile industry, the electronic industry and the office machine industry. In the beginning of the development mainly bigger systems were in operation. The trend currently converts to smaller systems with two to four machines, whereby the systems with replaceable machines prevail. The multitude of components of a flexible manufacturing system as well as the possibility of considering operational circumstances of the users make clear that these devices are customized for a specific application case. Step-by-step introduction of flexible manufacturing systems in production has proved very efficient, and the modular structure of present systems allows a continuous rise of the level of automation.

Flexible manufacturing systems operate either as autonomous devices or as modules of a computer-integrated production concept. Different control strategies have been developed in the field of flexible manufacturing systems; however, a hierarchical control structure with superior DNC control is a common pattern of implementation (Fig. 5.5).

Flexible manufacturing systems are becoming more and more integrated in superior, company-wide information systems on the basis of open

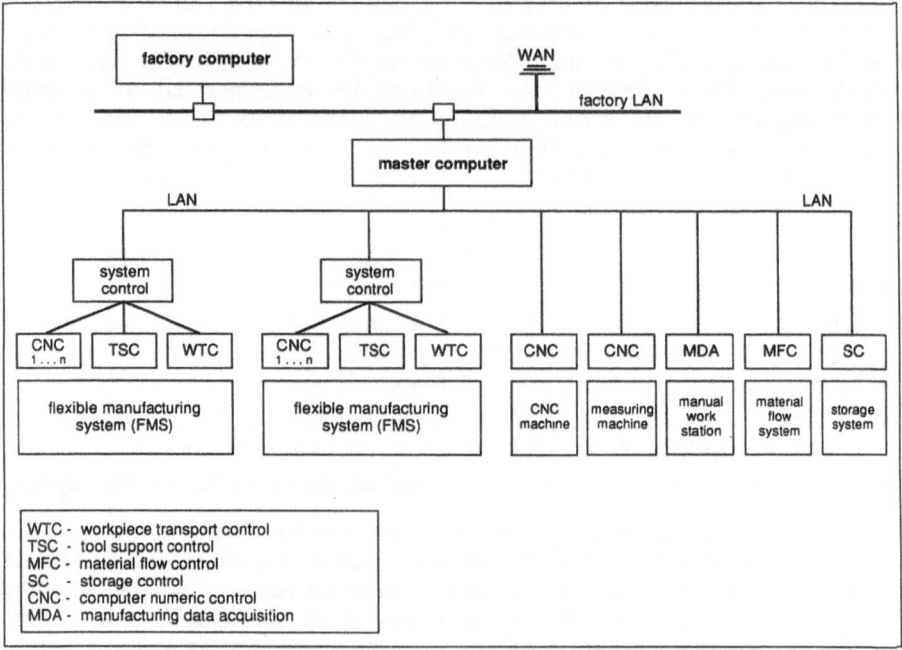

Fig. 5.5. Example of a flexible manufacturing system control.

local communication networks. Investigations show that today's flexible manufacturing systems in European companies are intended mostly to be integrated in the information system of the whole factory, so that the main demand of CIM – a free information flow – can be fulfilled to a high degree.

Due to their complex structure concerning the numbers of connections and the demand for a synchronicity of the material and information flow, flexible manufacturing systems are highly prone to technical and organizational failures; however, these can be minimized through reduction of the whole system complexity as well as through fast interchangeability of system units.

Working with complex technological systems is connected with higher requirements of knowledge and skills of the system leaders as well as of the maintenance personnel. The work on automated manufacturing facilities with automated workpiece and tool transport within and between machining centres, and their interaction with superior computer systems, requires especially:

- Organizational thinking for efficient utilization of the machining centres.
- Strategies to overcome deadlock periods quickly.
- Expertise for accurate communication and co-operation with maintenance technicians in case of interruption.

These points also emphasize the significance of improved methods in technical documentation, i.e. handbooks and leaders' guides to systems and plants. The need of fast availability of knowledge in order to control computer-integrated production systems is too important to be left to coincidence or to be handled as an additional occupation. The use of knowledge-based systems in the factory aims at knowledge supply and knowledge transfer, which places increasing significance on economic operation of complex technological systems through the use of appropriate human-independent knowledge storage.

5.2.3 Communication Networks

The realization of standardized interfaces and link protocols is a precondition for a continuous network of devices, computers and programs of different manufacturers. The physical connection is realized by an installation of local networks within the factory as well as by an establishment of supra-regional networks by postal services or other communication enterprises. With the help of these networks, internal and external communication as well as the procurement and processing of information will come about. The concept of the "paperless" factory can hardly be realized in the near future, but much information processing work can be carried out with direct access to electronically stored data and program functions.

Local area networks (LANs) allow a fast and safe information flow between computer systems which are linked by a common transmission channel within a limited geographical area. The definition of LANs includes transmission technology, medium, topology, connections or rather connectors and network access methods. As a response to this the Institute of Electrical and Electronic Engineers (IEEE) has standardized three different LAN types with different parameters, which enable a maximum network application in different enterprise departments. LANs for manufacturing as well as for other enterprise departments are chosen mainly according to application and costs.

Due to the often-used strategy of stepwise introduction in the field of computer-aided manufacturing systems, it is essential to pay attention to technological independence of components, which leads to the possibility of linking components from different product sources. The application of LANs on the basis of standardized cablings does not guarantee the possibility of free information flow, but represents an essential basis for it. Only a common communication architecture with unique proceedings, records and services makes communication possible using components from various sources.

A communication architecture is the sum of all rules in information exchange between communication partners. Besides other aspects the determination of communication functions, data forms and transmission technologies are important parts of it. Development of international standardized communication architecture sets the aim of free flow and of producer-independent manufacturing. The International Standards Organization (ISO) has worked out a framework in the form of a seven-layer reference model within the Open Systems Interconnection (OSI) Project and also established a number of single standards. Therefore it is possible to develop a communication architecture for specific applications.

Manufacturing Automation Protocol (MAP) is such an application-oriented communication architecture specified by General Motors (GM). MAP is mainly based on already existing ISO standards and takes especially into consideration the communication needs and marginal conditions of the manufacturing area, e.g. determinized response-time behaviour, high data security as well as often large geographical extension.

Many of the cell computers and cell programs used in current production are developed by different manufacturers, thus it is expected that the MAP standard will gain more and more significance in future. Although the range of suitable products is still limited, a stepwise implementation of MAP applications can already be effected by starting pilot projects in order to gain experience.

The development of the computer-integrated factory requires not only compatible networks and communication protocols, but also compatible interfaces. At the moment, efforts to standardize CAX interfaces include the areas of CAD data exchange (VDAFS, SET, IGES), product data (PDES, STEP), robot control (IRDATA), graphic processing (GKS, PHIGS) and manufacturing of standard parts (VDAPS).

The ESPRIT programme for CIM, supported by the European Commission, has founded the AMICE consortium consisting of 21 European companies in order to achieve further linking developments. The result of this work is ESPRIT CIM-OSA (CIM Open Systems Architecture), which includes a definition of a framework for CIM System Integration and the information technological support required.

5.3   Examples of Computer Integrated Factories

5.3.1   Integration of a Single Factory

Example 1: Machine Tool Company

The product programme of the company includes machine tools and work-stations as well as flexible manufacturing cells. The production is divided into components for metal removal tool manufacturing as well as planning and machining systems for sheet metal processing. The goal of the growing use and linkage of computer-aided design, manufacturing, assembly and material flow is on the one hand an increasing flexibility of production, and on the other an expansion of the daily working time up to a third shift with very few workers. In the field of sheet metal processing this concept with a constant information flow between the computer-aided design, NC programming, production planning and control (PPS) system, material handling system and machining units is already highly advanced.

The material flow in manufacturing is carried out by two automatic guided vehicles (AGVs). In the metal removal tool manufacturing, machines from different suppliers and different work-holding fixtures are used. Therefore high demands are put on the selection of vehicles during the period of planning. An important part of the CIM concept is a flexible manufacturing system consisting of two machining centres, one store for 20 pallets and two workstations. In this way, 70 different workpieces can be produced in a three-shift system. It was possible to reduce the throughput time to about 30%–40% by switching from conventional manufacturing to processing centres.

Two CNC manufacturing centres, equipped with automatic loading and unloading devices are used in sheet metal manufacturing. Furthermore, a CNC punching and nibbling machine as well as other processing stations, e.g. straightening machine, deburring device, bending-off press, robot-welder and manual welder go into action. The plant works in a two-shift system, whereas both the other processing centres operate in a three-shift system.

The planning level, consisting of the PPS and the computers for the NC programming, provides the master computers with the data required for the metal removal tool manufacturing and the data needed for the sheet metal part manufacturing as these are the production plan and NC programs (Fig. 5.6). These production master computers take over the co-ordination of three areas:

- The tool management which is controlled by an own computer system. It is possible to call in the status data of the most important tools at any time.

- The storage supervision and the material flow management which are also controlled by a dedicated computer system.

- The separate machines and systems of manufacturing. Some manufacturing devices are controlled by a special cell computer which is combined with the production master computer.

The master computers therefore are the central elements by which the

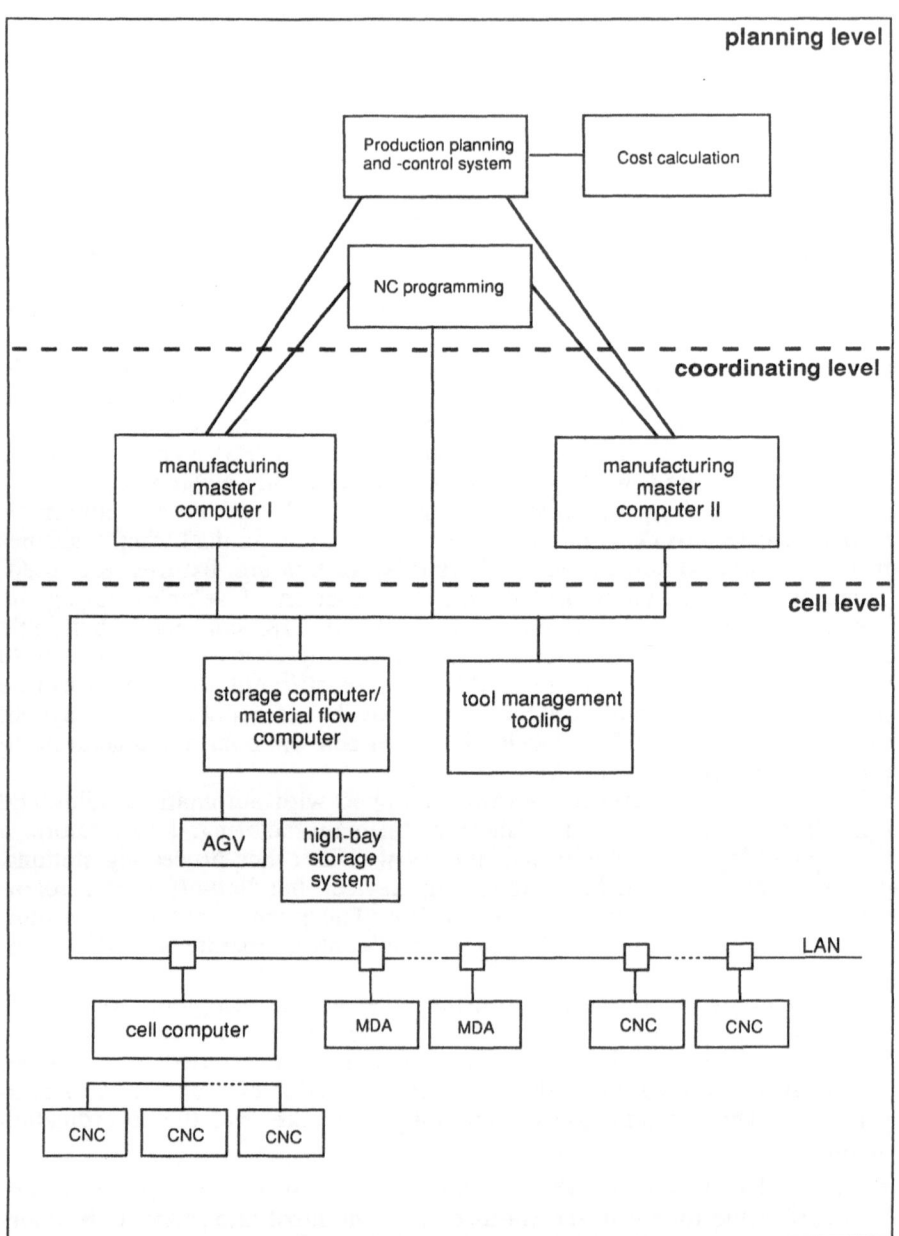

Fig. 5.6. System structure example 1.

production planning and control system are linked between the development system for the NC programs and the manufacturing equipment. As far as the manufacturing is concerned the main system takes over order information by the PPS system and controls the release of each order according to schedule. The distribution of orders to the machines takes place according to the optimal sequence following the criterion of minimal set-up time.

The production master computer processes the data of areas as tool management, material flow control and manufacturing and reports them back to the planning level, thus co-ordinating the supply of material and tools, of material flow and machine utilization. The control of the production master system registers operational data as processing time, manufactured quantity, beginning and end of the order processing. At conventional places of work this information is entered into the computer by factory data acquisition terminals. The PPS system brings order planning up to date on the basis of manually entered order feedback at these work places. The previous practical experiences with computer-integrated metal removal tool manufacturing and sheet metal part processing fulfilled expectations as far as higher machine availability, reduction of the throughput time, more reliable deadline planning, less material stock and better use of plants are concerned.

5.3.2   Integration of Overlapping Factories

Example 2: Machine Tool Company

The production of this enterprise essentially includes a CNC lathe and CNC turning machines as well as turning centres. The company produces at three different locations in the south of Germany. It is divided into an administration and a production department within the main building and two additional factories which are situated further away. The company is following the concept of flexibly automated production because of an increasing demand for special customer orders and faster delivery times.

The computer structure (Fig. 5.7) includes a central CAD/CAM computer providing information for the design, the resource and production planning and the technical order processing. Two additional central computers take over tasks of purchasing, deposition, storing, and of planning and controlling of manufacture by way of the PPS. This outline forms the basis for a gradual installation of partial solutions by information technology. Only this way was it possible to recognize and solve problem areas like those concerning interfaces. Thus the gradual installation and integration of individual CIM components could be initiated.

Each one of the three production locations is equipped with a production master computer. These three central computers are linked with each other, providing a linkage between production and the previous production planning and controlling. There is also a linkage between the PPS system and the departments sales and services. The linking of PPS and CAD is planned. Among other things it is a goal to make the parts list directly available for the production planning and control system. The CAQ system as well will interface with the PPS and via test plans and drawings also with the computer-aided planning system. Different numerically controlled manufacturing factories like lathes, machining stations, milling and grinding machines as well as pipe-bending and laser-engraving machines are connected to the computer with DNC tasks.

The computer-generated drawings and programs of design and develop-

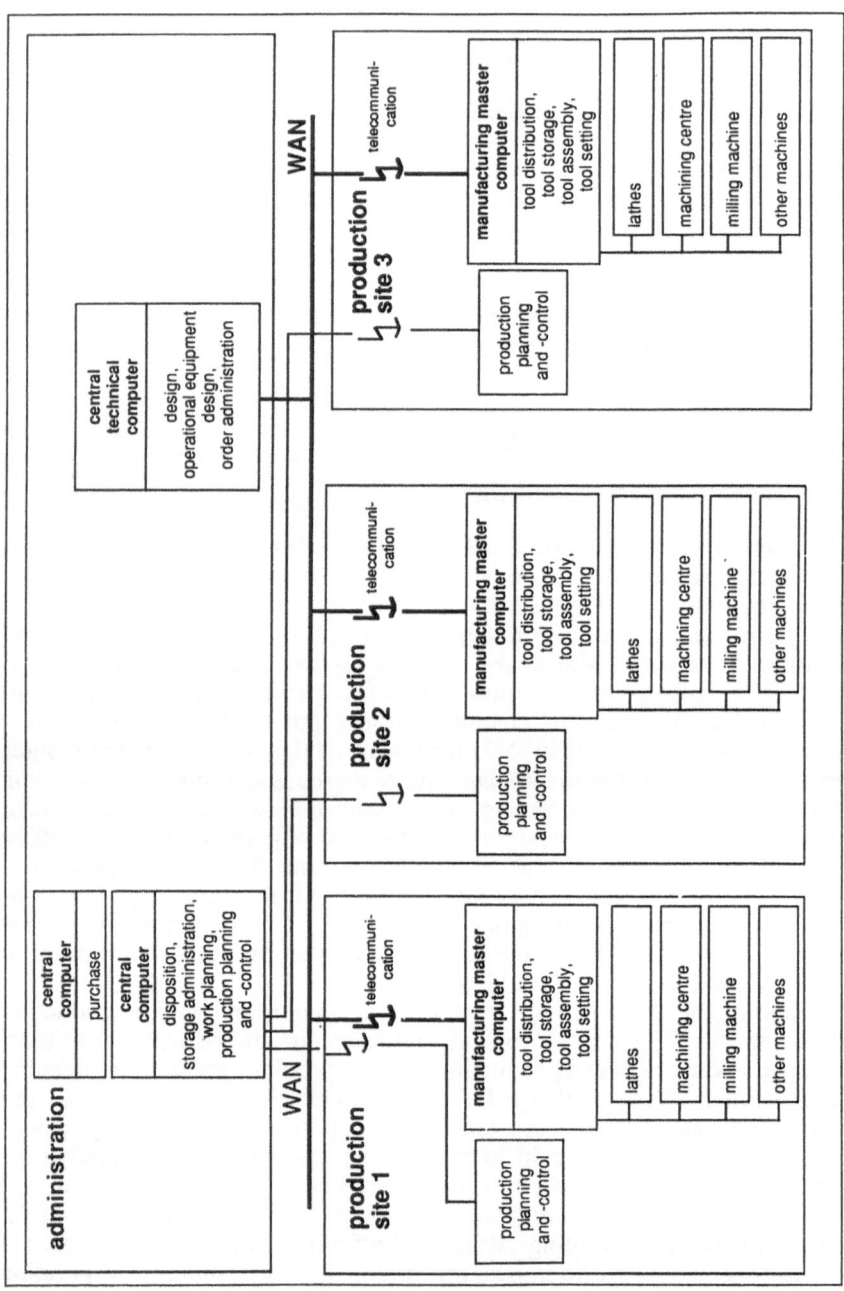

Fig. 5.7. System structure example 2.

ment are stored in a central computer. Because of the linkage of the central technical computer and the three computers within the factory this information is directly available in all production areas. Control programs can be loaded in situ by the computer. NC programming needs data which are already registered in CAD drawings. In order to make computer-integrated geometry data immediately available for NC programming, a CAD-neutral CAD/NC-linking element has been developed. The CAD drawing is stored as an APT or IGES data file and used for working drawings, chucking drawings or test plans. These drawings are directly retrievable by the computer throughout the company. Programmers produce neutral parts programs via the CAD/NC-linking element. They are stored in the processing computers of the three manufacturing locations.

This solution increases the profitability of a CAD system as double data acquisition is avoided, and therefore additional sources of error by data transmission are excluded to the greatest possible extent. In addition, the throughput time is considerably reduced. At the same time, NC programs can be optimized at an early stage by using the possibility of simulating the tool movements in order to avoid collisions.

The DNC computer supplies the numerically controlled manufacturing devices with the necessary NC programs. The DNC system is able to connect all NC machines of the respective production plant. A production master computer takes over the transmission of information, data and NC programs between the computer levels. In addition, this computer also takes over the tasks of program management, storage, and tool and operating data management. In order to guarantee the most accurate transmission of NC programs to the CNC controlled workstations, transcription of the stored original programs by back-transferred optimized NC programs is not possible automatically. They are administered in a special department, so that technical comparisons can be made to optimize the programs quickly and simply. All part programs which have been converted into NC programs by a postprocessor, as well as all NC programs which have been directly generated at the shop-floor level and back-transferred, are stored in a central record archive.

Example 3: An Electrical Equipment Company

The main activities of this company, which produces at three locations in southern Germany, consist of the production and the sale of pneumatic and electronic components, machine and control components, and electronic and compressed-air tools.

At the three locations, CIM cells have been installed to correspond to the increased demands concerning flexibility, productivity and quality. These partial CIM solutions are then linked to a computer network (Fig. 5.8). A central host computer in the main factory takes over the electronic data processing of the computer-integrated design, of order planning, material supply, stocktaking and of the planning of quality. The three factories are connected to the host computer by dedicated lines with terminals. The connected departments can always fall back on the actualized data via a company-wide communication network. The hierarchy of computers with the central host computer and the decentralized computers of the factories

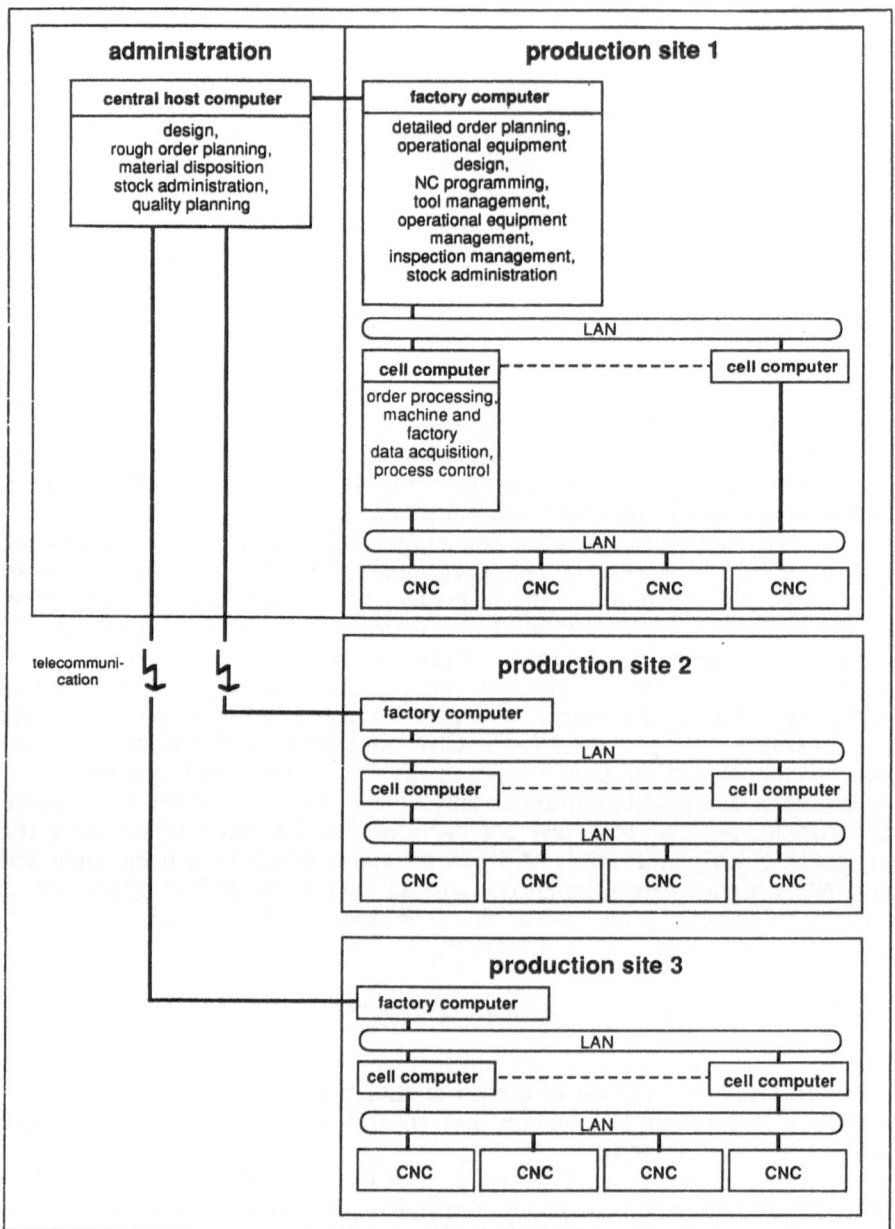

Fig. 5.8. System structure example 3.

provides a useful division of functions between the head office and the individual factories. The factory computers take over operational functions such as detailed order planning, capacity design, NC programming, tool management, capacity management, test and maintenance as well as storage order. The cell computers take over the tasks of order processing, machine and factory data acquisition and the process control in the factories on the process level.

In spite of the very different manufacturing structures within the three factories, standardized innovative solutions concerning partial functions in production and information engineering are developed for use throughout the company. The implementation of specific solutions is to be avoided in order to minimize expenditure on maintenance later on. Within the introductory concept it is intended to carry out and test the plans within smaller pilot projects. After such pilot projects have been carried out successfully they are implemented in the respective factories. Thus standardized CIM cells emerge which can be developed to enable comprehensive computer-integrated production throughout the enterprise.

Example 4: A Company in the Aircraft and Space Industry

The company develops, produces and sells products for air and space technology, military engineering, energy technology and medical science. The production is divided into five factories, two in the north and three in the south of Germany.

The company first planned to implement comprehensive computer-aided production technology in the mid 1970s. An adjustment of the structure of production was necessary because of the rapid development in aeronautics. The manufacturing of current products is clearly more complex, so that computer-integrated production is included in the further development of the factories.

Against the background of developments in aeronautics the use of new manufacturing technologies, as well as new materials like carbon-fibre-reinforced plastics, super plastic forms or new alloys such as aluminium–lithium, is planned. As far as cutting is concerned a high degree of automatization is the aim.

Each factory has its own production master computer (Fig. 5.9). A master computer at the main location includes a CAD/CAE component, whereas the subsidiary production master computer at the respective factory serves computer-aided design, computer-aided planning and computer-aided quality assurance. The computers are linked, so that design data produced at the main location can be processed further in other factories.

This computer structure also provides optimization of the total situation in the case of interruption of the manufacturing process within one factory. The master computer co-ordinates and controls the whole production process from material planning to shipping. The subsidiary production master computer serves the DNC computer as well as the tool and material system computer on the lowest level. The autonomously operating controls of the single machining and material flow systems also belong to this level. The DNC computer supplies the tool machines with necessary parts programs and is also in charge of the acquisition and evaluation of machine data.

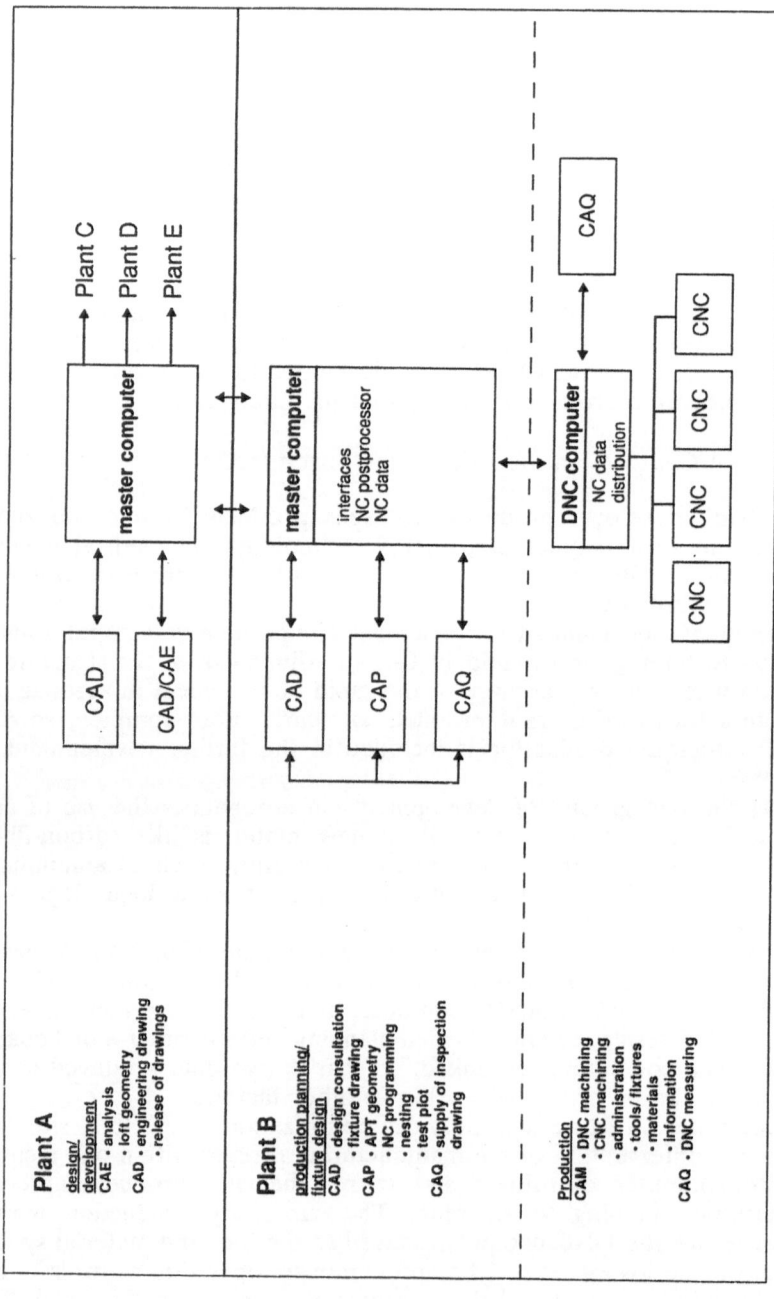

Fig. 5.9. System structure example 4.

Example 5: A Company in the Electrical and Computer Industry

The company produces drive units, steering and computer systems. The production plant for office systems is an international leader in its organization of manufacturing and output capacity. More than 1000 of such systems as display units, personal computers and multi-user computer systems leave the factory every day. A high degree of automatization in production and production logistics enables very short processing time and delivery periods. Delivery periods of less than 24 hours are possible, and small lots can be made.

An increase in production quantity demands a considerable extension of capacity. Two of the four planned production halls have already been put into operation. Both factories work autonomously, but they co-operate closely with each other. The goals pursued are particularly a considerable decrease in capacity, short-term delivery as well as the realization of flexible production. The use of knowledge-based systems in production is already at the planning stage.

Today the company has access to the resources of each connected computer via a network. This provides quick access to the production from various sides for production planning and control. The structure of the computer hierarchy is divided into planning, master, operation and processing levels, so it is possible to continue daily production even if the planning computer does not work. Factory data are transmitted with the help of electronic services of Deutsche Telekom like Telefax, Telex or Teletex. An electronic document exchange between the computers of different producers following the tentative standard of EDIFACT is a novelty. At the moment agreements with some important suppliers concerning electronic document exchange are being made. It is to be expected that electronic document transmission will be expanded in the future. Order processing is to be accelerated, information flow should be safer, double acquisition of the trade data avoided and costs reduced by introducing this system. The construction of an automated distribution centre which is to optimize the material and information flow within the plant and which is to guarantee better communication between producer and customer will be a further step.

5.4   Development of Organization and Human Capital in Computer-Integrated Production Systems

The entire benefit of the introduction of computer-integrated production can only be achieved if technology decisions are dealt with as part of the whole strategy of the company. It needs to be co-ordinated with all the other strategic goals. Only to a limited degree is traditional production management suitable for the computer-integrated factory of the future. Considering the rapid development in the technological and the economic environment, production management is not only expected to strive for productivity; its task today is rather to realize the potential advantages of competition in the new production technologies and to acquire possibilities of their realization under economical and personnel aspects.

The integration of automation islands will create network units and it will support the development of the entire factory towards an integrated information machine. The all-important characteristic of these operating cells is their flexibility that helps to overcome the interfaces. For successful use of computer-aided production equipment it is decisively important to optimize the given course and information flow. Therefore not only an investment in computer hardware and software is necessary but this process needs especially to be accompanied by qualified personnel. The data-driven factory is supported by the know-how of the employees, the knowledge, experience and qualifications and above all the creativity of the people involved.

With the new factory and its technology the structural change towards the tertiary sector will continue within the European labour scene. In the course of a decrease in manual labour and human-related manufacturing, activities like programming, supervising and controlling, repair and maintenance as well as creative activities like research, invention, planning and management are increasingly gaining importance.

Qualification is already today one of the key factors in the further development in production technology. Thereby the structural change will, on average, lead to higher qualification standards. There is a demand for more abstract, theoretical, systematic, organizational and planning work. Work in a future industrial society requires a high degree of cognitive education. Only someone who understands technology is able to control it.

The development of production technology cannot be separated from its economical and social environment. Its utilization calls for a high degree of social co-operation and communication. Thus the importance of a widespread, overlapping technical and social qualification is growing. The more production goals lead to manufacturing of highly complex quality products and require an extensive use of new technologies, the more a comprehensive scope of duties and a wider application of qualifications is required.

Qualification as a management task is not only a result from the pressure of the progressive specialization of professional abilities. The trend towards specialization seems to be embedded in a growing integration of manufacturing tasks that are beyond the traditional scope of the defined professional image and tend to include the entire factory (Fig. 5.10). This dynamic extension of professional fields concerning the cognitive decision and action process calls for an improvement of personal competence, thus a further education that is characterized by a generalization of specialized knowledge and skill.

According to the branch-overlapping mobility and flexibility of the professional qualification requirements, further education plays a central role today. No matter if it is organized within a company or supplied by external institutions, further education and training demand a concept integrating man and technology. The companies as the frame of reference must be regarded as socio-technical systems, where the use of new technologies without appropriate further education means a waste of resources. Therefore the increasing importance of professional education is a sign of the development in qualification that is connected with industrial innovation. The professions of the future will not only require academic

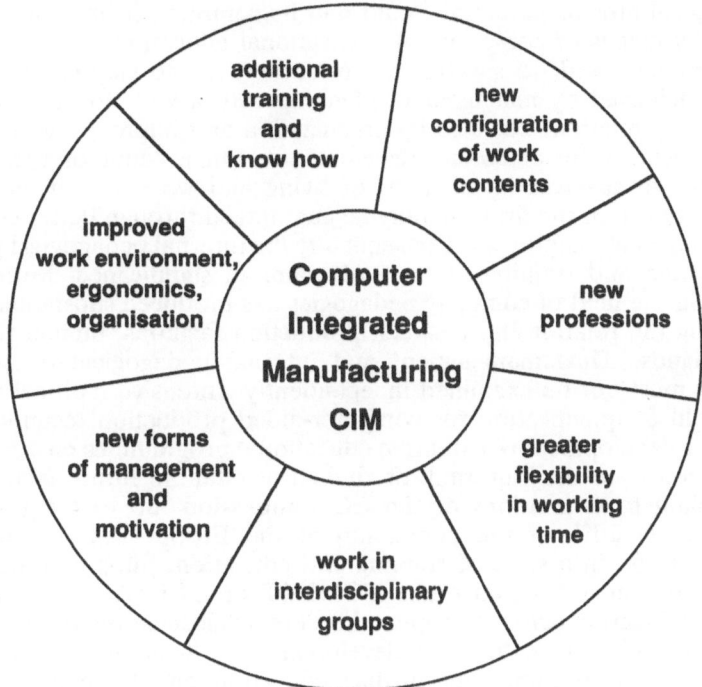

Fig. 5.10. Transformation of work enabled by CIM.

qualifications; people will develop towards a profession by vocational training and further education. That applies especially to today's phase of significant technological changes and economic challenges.

In such times there is a strong demand for capability and willingness for constant learning and flexible adaptation to changing situations. The result is an increasing demand for advanced training and further education as well as retraining measures. If deficits in qualification are not to be the obstacle to innovative structural change in the future, a complete system of further education needs to be built up very quickly. The urgency of such a requirement is also obvious by looking at the demand for education of the unemployed.

Investment consideration and the determination of the necessary changes in the qualification profile need to be started at least at the same time, in order to be able to use the planning stage to develop qualifications as necessary. But even constant adaptation of staff qualifications to the production structure cannot prevent them from being outdated by every technological innovation. Thus the necessity of qualifications is always seen in the shape of deficits. The result is a chronic delay in providing qualifications. This dilemma between an optimized qualification and a simultaneous reduction of its flexibility would result in a system of qualification potential that would reach beyond a mere technological competence with regard to special innovations. Thus it would become possible to anticipate future innovation requirements. In this way the technological growth potential would find its complement in a qualification strategy by means of comprehensive vocational education.

A company's goal, to survive in the market, i.e. to stay competitive, is not only achieved by management planning with a view to optimizing the input–output relation, but also by organization and improved utilization of human efficiency by personnel development. On account of the growing importance of sciences in all areas of living and working, an increase of productivity within the firm can only be guaranteed through better education and training of all employees. Consequently the internal pedagogical problem of promotion and training of the staff gains in significance. More recent research in the field of company pedagogics has produced substantial results concerning the relationship between production facilities, human resources and demands. Thus management and internal pedagogical questions or problems must not be examined independently. Intensive national research in the field of qualification for computer-aided production technologies as well as the development of extensive educational programmes on a European level demonstrate the high rank of "human capital" in future factories.

According to estimations of the EC commission, up to the year 2000, approximately 20% of the population of the European Community will permanently be in a state of training and education. Support programmes of the European Community like COMETT or DELTA are supposed to stimulate education across European borders by close co-operation between companies and universities, by developing new technological auxiliaries (equipment) for training and further education and by improving local educational opportunities. Furthermore, COMETT will very much stimulate the exchange of students across borders within Europe. Therefore future managers will have the opportunity to become acquainted with economic, social and cultural situations as well as the mentality and characteristics of different European countries.

5.5 Outlook

The slogan "factory of the future" is to be understood as a fundamental structural change of the factory rather than a utopian vision. This change is characterized by an evolutionary development of the stages of automation. Although flexible manufacturing units and computer-integrated production systems open up considerable potential for a rationalization of the entire production process, these chances are not being adequately used in business. In competition and especially on an international level, a company can only survive in future if efficient manufacturing equipment is accompanied by an organization and staff equally qualified.

The establishment of the Single Market and the democratization of Eastern Europe will play an important role in the development of close economic and social bonds within Europe. The European industrial enterprises are preparing for this change in surroundings by including new markets for acquisition and distribution in their planning. Already, co-operation across European borders indicates a trend towards closer co-operation between the companies.

International co-operation, the merging of companies, production compounds and the necessity of the presence on international markets require increasingly efficient electronic networks and transmission techniques. Only in this way can a fast exchange of information over great distances be guaranteed for the future. Today the major part of European data exchange is still handled over telephone and telegraph networks that are operated by different national postal administrations and communication authorities with their own suppliers, systems, services and standards.

The Deutsche Telekom for example, the supplier of the public data network in Germany makes two transmitting media available for coupling of LANs over a long distance. The datex-P/X.25-network and the more powerful direct call network are the WAN connections mainly used for data traffic at the moment. These networks are already completed by a broadband communication network (ISDN – Integrated Services Digital Network) that uses available telephone networks but considerably extends their capacity with the help of digitalization. This performance will be improved with the intended use of high-speed networks using fibreglass. In 1992 ISDN started everywhere in Europe. Although decentralized manufacturing structures are still at the beginning of a further development, technology already provides the necessary methods and instruments for a close communicative and logistic connection of factories. Especially in the area of the European automobile industry with its demands for just-in-time-delivery the direct information flow between suppliers and the assembly plants is highly advanced. An effective data exchange among different factories requires an agreement on a data format that can be used by various computer systems.

Within the EDI (Electronic Data Interchange) different standards for business data have crystallized within Europe. The ODETTE standard (Organization for Data Exchange by Tele-Transmission in Europe) is of significant importance; it is a European development that is especially promoted by the European automobile industry. In addition, further standards promoted by different branches have been established, for example CEFIC for the chemical industry and EDIFICE for the electrical industry. These efforts towards standardization aim at a faster document-free data exchange among manufacturing and service enterprises. It is to be expected that the electronic data exchange will even be used by a number of small and medium-sized enterprises. Besides the acceleration of the data flow there is also a need for an increase in the capacity of the available transport system. A growing flow of goods and a high transportation frequency, especially for just-in-time supply, today limits the capacity of many European traffic roads. Thus possibilities for further development in decentralized manufacturing structures need to be examined, particularly under the aspect of future logistic systems.

Further Reading

Alschweig E. Zwischenbetriebliche Kommunikation in einem Großkonzern. CIM-Management, München 1989; 5 (3): 31–33

Badham R, Schallock B. Human factors in CIM: a human-centered perspective from Europe. Int J Hum Factors Manuf 1991; 2: 121–141

Burstein D. Weltmacht Europa. Wilhelm Heyne Verlag, München, 1991

Eversheim W, Bußmann, J, Rathjen C, Wiegershaus U. Strategien zur integrierten Produktion. VDI-Z, Düsseldorf 1989; 131 (9): 65–68

Ewers H-J, Becker C, Fritsch M. Wirkungen des Einsatzes computergestützter Techniken in Industriebetrieben. Berlin/New York: Walter de Gruyter 1990

FAST-Gruppe (Hrsg.) Die Zukunft Europas – Gestaltung durch Innovationen. Springer, Berlin, Heidelberg, New York, 1987

Hanewinckel F, Küspert K. Integration durch objektorientierte Datenbank. In: VDI-Z, Düsseldorf 1990; 132 (3): 50–57

Herrmann R. Netzwerkmanagement. Des Elektron Markt Techn 1991; 6: 34–45

Herter J. Qualifizierung für flexible Fertigungssysteme. In: Spur G (Hrsg): Produktionstechnik–Berlin, Band 89. Hanser-Verlag, München, 1991

Institut für Systemtechnik und Innovationsforschung (ISI) Stand und Aussichten der Fertigungsautomation in der Bundesrepublik Deutschland. Endbericht zum Forschungsauftrag Nr. 22/87 I an das Bundesministerium für Wirtschaft. Karlsruhe: 1989

Krallmann H. (Hrsg.) Zwischenbetriebliche Integration (ZBI). CIM-Manag München 1989; 5 (3): 3

Milberg J. (Hrsg.) Wettbewerbsvorteile durch Integration in Produktionsunternehmen. Referate des Münchener Kolloquiums '88. Springer, Berlin, Heidelberg, New York, 1988

Nijkamp P, Reichman S, Wegener M. Euromobile: transport, communications and mobility in Europe. Aldershot, Brookfield, Hong Kong, Singapore, Sydney, 1990

Schulz-Wild R, Nuber C, Rehberg F, Schmierl K. An der Schwelle zu CIM. Verlag TÜV-Rheinland, Köln, 1989

Shah R. Flexible Fertigungssysteme in Europa: Erfahrungen der Anwender. In: VDI-Z, Düsseldorf 1991; 133 (6): 16–30

Specht D. Wissensbasierte Systeme im Produktionsbetrieb. Hanser-Verlag, München Wien, 1989

Spur G. Entwicklungstendenzen rechnerintegrierter Fabrikstrukturen in Europa. In: ICMA, Vorträge zum internationalen Kongress für Metallbearbeitung, Hannover, 1985, pp 6–14

Spur G. Einführungsstrategie zu CIM. In: VDI-Z, Düsseldorf: 1988; 130 (10): 12–14

Spur G. Unternehmensführung in der zukünftigen Industriegesellschaft. In: Spur G. (Ltg.): Produktionstechnisches Kolloquium, Berlin 1989, Vorträge, Berlin, 1989, pp 5–15

Spur G, Specht D. Flexibilisierung der Produktionstechnik und Auswirkungen auf die Arbeitsinhalte. Z Arbeitswiss, Köln; 1987; 41 (4): 207–11

Spur G, Herter J, Zurlino F. Qualifizierung für flexible Fertigungssysteme durch die Herstellerunternehmen. Z wirtschaftliche Fertigung Automatisierung München 85 (1990) 11, 605–608

Spur G, Specht D, Zurlino F. Zeitorientiertes Fabrikmanagement als Erfolgsfaktor im Innovationswettbewerb. Planung Prod Heiden 1991; 39: 12, 19–25

Tritremmel W. Fabrik 2000: Die Zukunft liegt im Vernetzen. In: Manag Z Zürich 1989; 58 (5): 55–58

Tröndle K, Weckerle E. Austausch von CAD-Daten zwischen Unternehmen. VDI-Z Düsseldorf 1989; 131 (3): 12–16

Wildemann H, Westkämper E. (Hrsg) Fabrikstrukturierung Europa '92. gfmt-Verlag, München, 1989

6 Flexible Computer-Integrated Manufacturing Structure of Global Network Type – Case Study in Japan

T. Itoh

Special attention must be paid to both the purpose and the origins of the flexible computer-integrated manufacturing structure (FCIMS) of global network type systems. Manufacturing systems have a number of key factors in their design principles, function and performance which are characterized by the cultural background of their origins. It is difficult, as a result, to obtain an overall view of FCIMS with the limited perspective of a case study conducted in one country.

This chapter, however, presents a single view of FCIMS of global network type by examination of video tape recorder production by Hitachi Ltd. in Japan. This is an example of the development of a flexible manufacturing control system which was developed after considering the questions "What should a human being do?" and "What should a machine do?". The answers were used to optimize the design for both economy and overall performance.

A local network consisting of a Hitachi Workstation (2050/32) and 14 independent personal computers (B16/MXII), was connected to a Hitachi M-280D mainframe computer.

The most advanced feature of the system was an interactive correcting system for production planning which controlled factors such as:

- Production quantity
- Parts delivery instructions
- Production progress management
- Supplementary ordering of damaged parts

These are systematically machine-processed to eliminate complex human handling. The system is applied to various of Hitachi's overseas operations and results in reduced manpower requirements in the factories' production control departments.

6.1 Worldwide Strategy of Video Tape Recorder (VTR) Business

Having established the basic production methodology in Japan, Hitachi commenced its European production in Germany in 1983, in Wales in 1985, and followed into Malaysia in 1990. Alongside expanding local production, each plant increased the local contribution of its own production by importing processes from the Japanese parent factory. The first factory in Germany started as a simple final assembly operation in 1983, but expanded the scope of the local processes to include high precision machining as well as the assembly of aluminium cylinders in 1987. The assembly of electrical printed circuit boards, which hitherto had taken place at the Japanese plant was gradually moved to both the Welsh and Malaysian factories.

Shifting the manufacturing processes overseas directly affects the complexity of production order management and process management, particularly during the transitions.

All parts described by the design engineer at the completion of a design have individual part numbers and subassembly numbers at various stages of the assembly process. Some parts need to be transferred to overseas factories as individual parts, whereas others are transported as either subassemblies or on the larger scale of master subassemblies. The precise scale clearly depends on the model in construction and the form of production. This is analogous to a level of file management.

The level of parts to be shipped is indicated by the parent company's master computer, located in Japan. The resulting components are shipped to local factories, as illustrated in Fig. 6.1. Hitachi's standard procedure for VTR production is as follows. On completion of the engineering design, a full description of its parts list is attached to the set of drawings. The list comprises:

1. Part number and subassembly number through various stages
2. Parts description
3. Shop indication
4. Quantity
5. Drawing number
6. Remarks

The "shop indication" defines which workshop is to perform any particular operation.

The core component of a VTR is the cylinder, sometimes known as the scanner. The cylinder is a sophisticated precision machined component comprising a number of high accuracy miniature subcomponents which assure rotational stability at the heart of the VTR. This is the single most expensive and delicate component which is referred to as the nucleus of the VTR, illustrated in Fig. 6.2.

As a result the in-process inventory, or waste caused in the cylinder workshop, is crucial as it affects all other shop processes. The total cost of the factory's production is therefore highly dependent on the performance of the cylinder workshop. Rationalization of the whole process in a factory

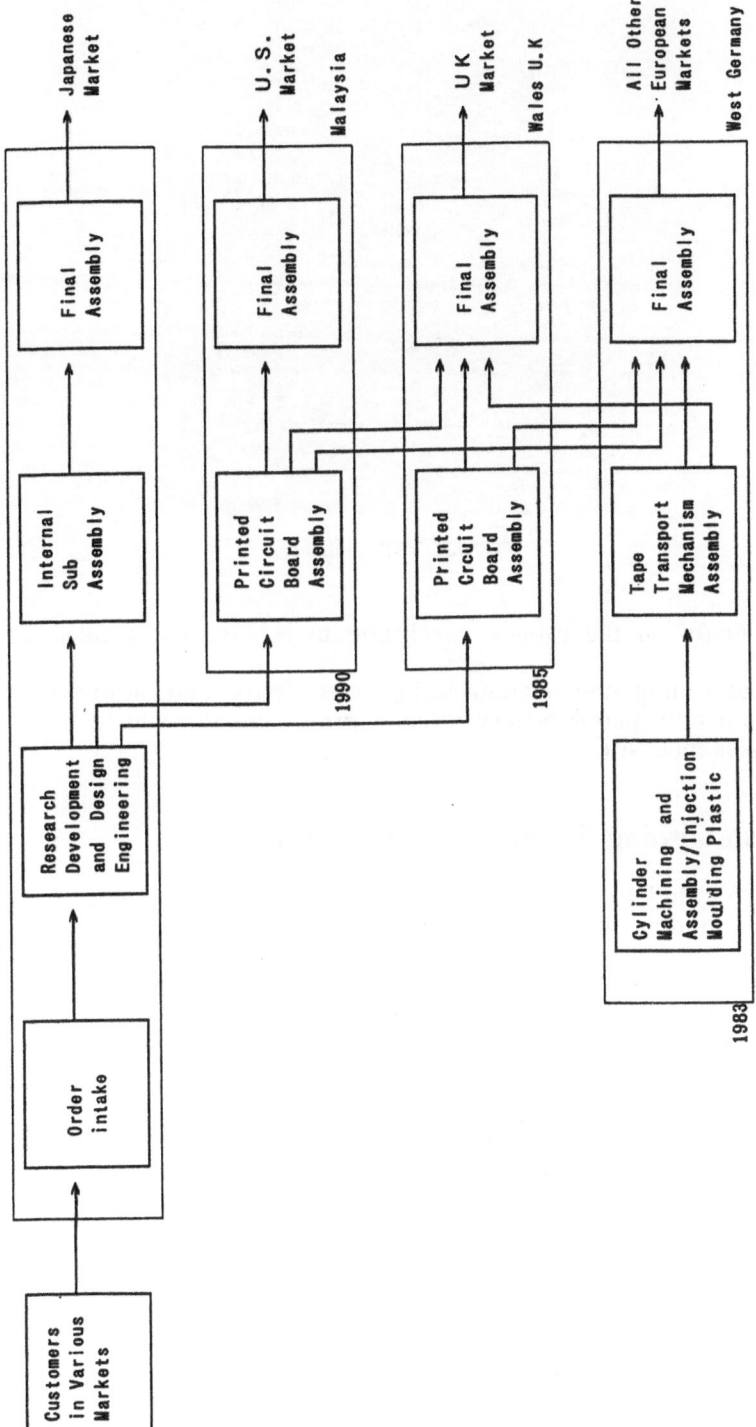

Fig. 6.1. Standard procedure of VTR strategy.

Fig. 6.2. VTR cylinder.

focuses firstly on the cylinder workshop to achieve the greatest overall efficiency.

Computer-integrated manufacturing was firstly and most effectively established in the parent factory in Japan, prior to being gradually transferred to overseas factories.

6.2 Establishing the Production Master Plan

The video cylinder is a self-contained component in a VTR. Its production is shown in diagrammatic form in Fig. 6.3. Its main components are:

• Video head assembly
• Upper cylinder assembly
• Lower cylinder assembly

The video head assembly is a tiny metal-based component which carries a sesame-seed-sized video head made of ferrite. It is built with very specific and advanced techniques, and takes a long time to produce.

The upper and lower cylinder assemblies are largely aluminium components with accurately controlled roundness, roughness and centring. The production process requires special attention to both machining and quality control. Inventory management is thus minimized by careful balance of head and cylinder manufacturing.

6.2.1 Analysis of the Cylinder Workshop

1. Because the cylinder is made from extremely fragile precision components, there is a serious risk of inducing failures during production. The next

stage process is the Transport Mechanism Production Workshop, which changes its production model at least six times per month. To maintain continuous operation downstream there is a psychological tendency to retain ample stock in the cylinder workshop, thus resulting in an extended production period and waste.

2. Parts are controlled by two procedures: the job number and the individual part number. These procedures require complicated inventory controls.

3. Determining the production quantity in terms of completed cylinder assemblies by counting the small individual components is time consuming, and requires large numbers of calculations. There is a significant risk of loss of parts.

4. To achieve job orders at each process station, the results of production are examined in conjunction with the ordered production schedule. This must be undertaken at all stations in a short period to obtain a smooth production run.

5. Manpower calculations require a lot of people to manage all requests, such as quantity levelling and the frequent schedule changes.

6.2.2   Job Number Management and Part Number Management

The order for VTR production is made by a job order which specifies quantities and the delivery date (job completion date). The job order has a specific number, which is referred to as the "job number" in Hitachi.

In the motor industry, many different models of car use the same kind of engine, and any one car may be powered by different types of engine. The same concept applies to VTR, as many models of VTR share the same cylinder. The same cylinder is used in a variety of VTR models, so a separate job order number for cylinders is used. Thus the production shop may produce larger numbers of a given cylinder type to improve productivity and efficiency, resulting in lower component costs.

Cylinders, like car engines, share common components with other cylinders. This commonly becomes more apparent the further upstream we go in the production process to the basic components.

At a production site, there is therefore a tendency to simplify production control by bundling parts with the same number from different jobs. Control is then by the part number of the cylinder rather than the job number of the VTR.

When a single workshop uses both job number control and part number control, there is a tendency to order excess parts. This ensures that each workstation is stocked with ample parts for long term production, and results in a lack of quantity control. It is also possible to omit to order small parts as a result of ignorance of the quantity required on a job order form.

Employing computers for these tasks avoids both problems.

6.2.3   Defining the Production Schedule

After accumulating customer orders a VTR production order is issued model by model, allocating production to the various lines. The subassembly (the tape transport mechanism) and its subcomponent (the cylinder) are bundled to form larger groups which may be common to more than two different mechanisms or VTR models. Separate cylinder production orders are generated, which are decided by the quantity of required VTRs. In general, the downstream process controls the upstream process: in this respect the VTR manufacturing is identical to any other manufacturing activity.

The automatic production planner in this CIM system has an interactive correction feature which handles demand and schedule changes downstream, as well as smoothing workshop load. The idea of using the downstream quantity demand to control upstream production rates is called material requirements planning (MRP) in this chapter.

6.2.4   Background of MRP Procedure Implementation

In general:

- At higher levels in the stream, more component commonality is found.
- Common parts are more often controlled by part number rather than job number.
- Using part number control achieves greater production schedule flexibility, and thus results in fewer control data overall.

These factors lead the system to make orders automatically for upstream components based on downstream requests as illustrated in Fig. 6.3. The production sequence is:

1. Cylinder final assembly line
2. Upper cylinder subassembly line
3. Lower cylinder subassembly line

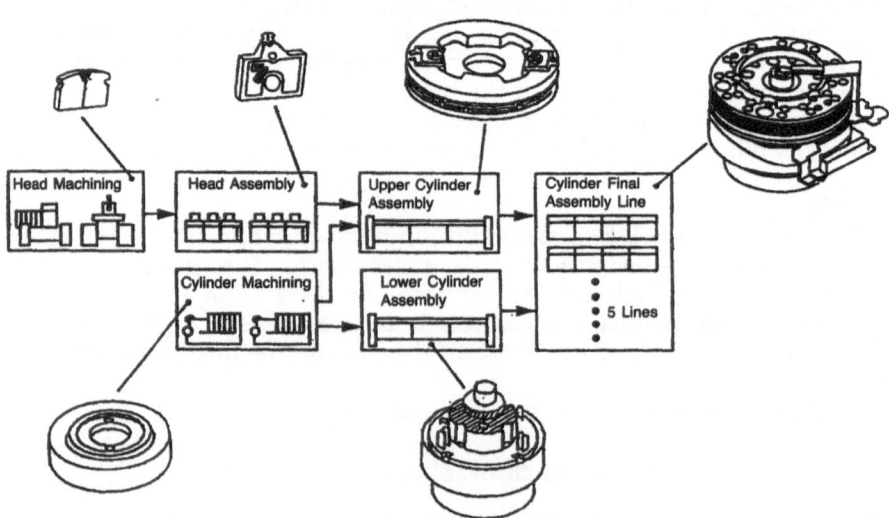

Fig. 6.3. Head cylinder shop process.

4. Cylinder precision machining shop
5. Head assembly shop
6. Head machine processing shop

The time taken to compute a parts request is fairly long, so a 32-bit workstation is used to handle the production scheduling. This is a Hitachi workstation (2050/32) running a database written in the C programming language to maximize speed.

6.2.5   Interactive Plan Correction

When operating under normal conditions, about five or six schedule changes in the production schedule for mechanism production affect upstream cylinder production. The corrective work, if done by hand, would take around 12 person days. Fine changes in the schedules are quite normal after the broad outline of the production lines has been altered. This tuning is required because the line may have insufficient equipment or labour to fulfil the revised plan. All these conditions require changes to the established plan.

By correcting the original MRP, it is possible to achieve:

1. Even distribution of workload per line.
2. The manner in which the adjustment of line capability affects production.

Drastic changes in downstream mechanisms may nevertheless require complete recalculation of the upstream cylinder plan, although small changes may be accommodated by the interactive correction facility.

6.2.6   Other Features

Many other cases require long recalculations even following a new MRP and interactive corrections to achieve an even workload. For example, if the production quantity of a particular product on a line has been changed, the total day's production, which includes the production from all other lines has to be changed. The system has then to check that the resulting production quantity will meet the downstream demand. These calculations are all handled by the workstations described above.

6.3   Production Ordering

The primary production planning sheet is a "Gantt" chart which may be interpreted to ascertain, for example what to make and in what quantities. The resulting order should contain a parts delivery instruction to the warehouse to obtain the required parts for delivery to the production line at the required times.

6.3.1   Work Load per Product

Smooth production rates are ideally obtained if each production line performs as requested in the production planning process. However, certain problems may jeopardize this state:

- Equipment malfunction
- Workers' sick leave
- Delivery problems with small parts
- Various other technical problems

Daily monitoring of production progress is required on each and every line in the production schedule.

6.3.2 Control and Delivery Instructions of Small Parts

Small parts tend to be delivered to production lines in enormous quantities because of their size. It is important to keep track of the quantities because of the high inventory cost, and also to ensure that they are not overlooked because their absence may hold up production. Information on the small parts is gathered, and their flow through the warehouse is monitored. The system has to be able automatically to provide status information about these parts, such as:

- Stored in warehouse?
- Present in production line?
- Which line?
- Job already completed and goods in finished warehouse?

Cylinder production is carried out at several of Hitachi's Japanese factories as well as their German plant. The system is designed so that the parts are delivered matched with the production schedule generated originally at their Japanese parent factory.

6.3.3 Data Collection from Production Result and Progress Control

The small components are, owing to their size and shape, difficult to incorporate into a bar code reading system. If the bar code system were usable there would still be many difficulties including the method of readout. As a result, manual data entry is used with provision for possible migration to a future automatic reading system. The manual system is designed to allow data to be sent at any random time for maximum flexibility.

Progress control is defined as a system which compares the plan on a daily basis to produce figures of the perturbations from the required figures.

6.3.4 System Structure

The system comprises four stations:

- Production planning and scheduling station
- Production order station
- Parts control station
- Results collecting station

These are connected to the Hitachi mainframe computer (M-280), as illustrated in Fig. 6.4. The group hardware comprises a Hitachi workstation (2050/32) and 14 Hitachi personal computers (B16/MXII).

The production planning and scheduling stations use the 2050/32, owing to the calculation load placed on it. It is located in the planner's office, some distance away from the production site. In contrast, a production order station is located at each production site, as follows, for ready access by the line foremen:

- Video head production shop
- Cylinder machining shop
- Cylinder assembly shop

Parts control stations are located in both the production control office and the parts warehouse, whilst results collecting stations are in each required workshop, as illustrated in Table 6.1.

The apparatus is interconnected as a network system using Field-net 600 between the 2050/32 and the B16/MXII, and Microsoft Corporation MS-NETWORKS™ between several B16/MXIIs.

Personal computers (B16/MXII) are used extensively for production ordering and parts control. Data communication between the B16/MXIIs is effected by MS-NETWORKS™ transfer speed 10 Mbit s^{-1}. This network provides for common data files, which enable data unification and selection. These factors eliminate unnecessary data transfer and result in efficient transmission. Using MS-NETWORKS™ significantly limits costs by reducing the development of communication software.

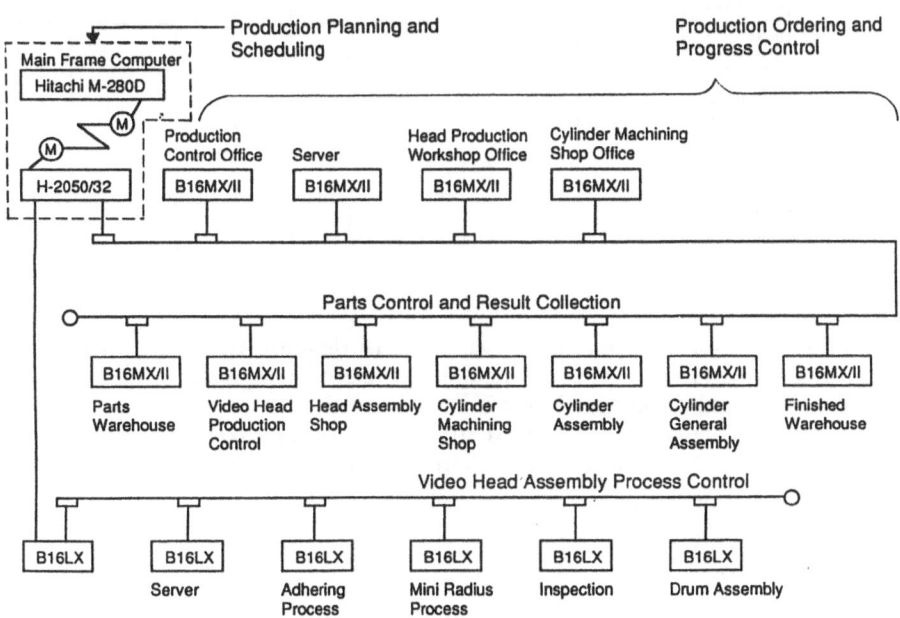

Fig. 6.4. System structure.

Table 6.1. Set-up locations and functions of personal computer

Location		Production Planning	Production Order		Parts Delivery Order		Parts Inventory		Parts ship date	Production Progress	Wasted Parts
			Plan	Disp.	Plan	Disp.	Calu.	Disp.			
Production Schedule	Production Control Office	o		o							
Production Order	Production Control Office 1		o	o						o	
	Production Control Office 2		o	o						o	
	Production Control Office 3		o	o				o		o	
Parts Control	Prod. Control Office			o	o	o		o	o	~	
	Parts Warehouse			o	o	o			o	o	
Result Collection	Video Head Production Control			o						o	
	Parts Warehouse			o							
	Machining Shop Office			o							
	Upper Cylinder Machining			o							
	Lower Cylinder Machining			o							
	Head Assembly			o							
	Upper Cylinder Assembly			o			o				
	Lower Cylinder assembly			o			o				
	Cylinder Final assembly			o			o				

The C programming language was used to develop the MRP production planning procedures as it could achieve efficient processing. Unify Corporation's relational database, UNIFY™ was used.

The BASIC language was used for certain other programming tasks to enable simple expansion, maintenance and faster development speed:

- Interactive correction
- Production job ordering
- Parts control
- Result control

Table 6.2 lists the number of steps in each functional process.

Table 6.2. Functional process and steps

Function	Steps
MRP Production Plan	3.0k
Interactive Correction	4.5k
Production Ordering	2.5k
Small Parts Control	1.5k
Result Collection	1.0k
Edit Transmit/Receive Data	3.0k
Editor	2.5k

6.3.5   Description of the Function of Production Planning

Defining the MRP Production Schedule

The input to this process is data detailing a four-month daily production schedule which shows the completion dates of downstream mechanism production based on the mechanism schedule. The schedule for upstream cylinder production is derived in the following sequence:

- Cylinder final assembly
- Upper cylinder assembly
- Lower cylinder assembly
- Cylinder machining (precision machining)
- Head assembly

The quantity of parts required from a workshop is calculated by examining the number of days required for production based on downstream production quantities. In the case of parts held in stock, the required net quantity for manufacture is the difference between the planned quantity and the inventory. The process breakdown in sequence is:

1. The four-month daily production schedule is input via a keyboard.
2. After considering the necessary production duration and the calendar, the downstream completion date for upstream cylinders is calculated and input.
3. The total cylinder production is computed from the mechanism job number.
4. Comparing the mechanism job order quantity and the result of 3, the planned quantity is modified until the figures match.
5. The quantity of a given cylinder type is decided.
6. The required quantities to be produced by each line are allocated.

This procedure is applied to each upstream production phase to decide

- Upper cylinder assembly

- Lower cylinder assembly
- Cylinder mechanism shop
- Head assembly

Steps 2 and 5 define the production schedule for each respective workshop. The data used to determine these processes are complex and are used by several stations. The information includes:

- Parts structure data
- Information on individual parts
- Calendar
- Information about the assembly line

To manage these data effectively, the relational database, UNIFY™ is employed.

Interactive Correction of Production Plan

The production plan explained above is based on data concerning downstream processes only. The peaks and troughs of the workload are not smoothed at this stage. To achieve constant efficiency and performance for a given production line it is necessary to have a constant production workload. The plan based on MRP should be tuned to maximize the performance of any particular line.

Figure 6.5 illustrates the interactive correction display. This example represents the production plan for a cylinder final assembly selected from various assembly lines. The following describes the keys listed in the diagram:

F1. Daily production change to replace the "rate of production" parameter

F2. Commencement date for change

F3. Quantity replacement for certain date

F4. Shift production of a particular quantity of given part from one line to another

F5. Delete a day: all production on that date, and the date itself to be erased

F6. Insert a day: recover deleted date

F7. Work day: change a planned holiday back to an active work day

F8. Select production line: swap to another line from the current

F9. Delivery quantity check: ascertain whether the planned quantity satisfies the downstream process

F10. Sorting key: reorder the part numbers to alphanumeric order

F11. List output: list by production line or by workshop total

F12. Cancel selected function: cancel interactive scheduling scheme or exit

Line : C
Line : B
Line : A

[Cylinder Final Assembly Production Schedule]

Model	20/May	21/May	22/May	23/May	24/May	25/May	26/May	27/May	28/May	29/May
100000	1,200	1,350	1,200	1,200	1,700	1,700	1,800	1,800	1,800	1,800
A-1	1,200	2,550	3,750	4,950	6,650	8,350	10,150	11,950	13,750	15,550
200000	300	300	300	300	300	300	300	300	300	300
A-2	300	600	900	1,200	1,500	1,800	2,100	2,400	2,700	3,000
300000	0	0	0	0	0	0	800	800	800	800
A-3	0	0	0	0	0	0	800	1,600	2,400	3,200
400000	0	0	0	0	300	300	300	0	0	0
A-4	0	0	0	0	300	600	900	900	900	900
500000	0	0	0	0	0	100	100	100	100	400
A-5	0	0	0	0	0	100	200	300	400	800
Total	1,500	1,650	1,500	1,500	2,300	2,400	3,300	3,000	3,000	3,300

F 1 P/D Quantity Change	F 2 Start Date	F 3 Quantity
F 4 Line Shift	F 5 Date Deletion	F 4 Date Addition
F 7 Workday	F 8 Select Line	F 9 Delivery Date
F 1 0 Sorting	F 1 1 List Quantity	F 1 2 Cancel

Cursor ↑ ↓ → ←   Scroll ⇑ ⇓ ⇔ ⇄   [End]   [Help]

Fig. 6.5. Interactive correction display.

6.4 Parts Management

The parts control station consists of two personal computers placed in the production control office and the parts warehouse. Small parts are managed by these two computers which manage the ordering of parts for delivery and record deliveries. Schedule production of

- Cylinder assembly
- Upper cylinder assembly
- Lower cylinder assembly

Thus the required quantity of small parts is calculated. Each component's required delivery quantity is calculated by comparing the accumulated deliveries against the planned number. The number to be ordered is arrived at by examining:

- Delivery accumulated request
- Delivery accumulated
- Failure quantity in process

Figure 6.6 shows an example of the daily delivery record. Further inputs may be made at any time.

Inventory control and discarding are managed as follows. Physical counting, including assessing the number of defects is carried out on a monthly basis at Hitachi. Figure 6.7 shows the production progress display and the number of rejected parts.

```
           Process : Upper Cylinder Assembly

      Date   Aug.9
```

No	Nomenclature	P#	Delivered	Failure
1	Shaft	100000	345	1
2	Pressure Metal	200000	1520	0
3	Bearing	300000	162	3
4	Motor	40000C	510	10
5	Motor Y	500000	412	2
6	Disc	600000	80	2
7	Transformer S	700000	475	5
8	Transformer R	800000	200	10

```
      F1   Past Record    F3   Delete    F16  End   ↑↓ Scroll
```

Fig. 6.6. Upper cylinder assembly parts delivery record.

[Progress]

No.	Process		Planned	Result	Balance	Achievement	Delayed	Ahead
1	Precision Machining		6,520	6,610	90	101.4%	−20	110
2	Head		5,112	5,842	730	114.3%	−231	961
3	Upper Cylinder	VTR	5,005	4,490	−1,315	77.3%	−1,345	30
4	'	Movie	4,650	4,422	−228	95.1%	−622	394
5	Lower Cylinder	VTR	7,410	7,512	102	101.4%	−52	154
6	'	Movie	6,850	5,970	−880	87.2%	−1,022	142
7	'	CKD	5,631	5,763	132	102.3%	−104	236
8	Final Line	A	8,055	7,460	−595	92.6%	−942	347
9	'	B	7,630	8,123	493	106.5%	−853	1,346
10	'	C8	6,595	6,410	−185	97.2%	−256	71
11	'	C4	9,851	10,082	231	102.3%	−45	276
12	Service Parts		6,807	5,912	−895	86.9%	−1,245	350
13	Final Line	H	7,430	8,033	603	108.1%	−90	702

[F16] End

Fig. 6.7. Production progress display.

The result control station displays the production order quantity, the result of the order, and the work in process inventory condition for small parts. Display selection is by function key:

F1. Past result display
F2. Copy
F3. Delete
F16. Quit

6.5 Progress Management

Since the production plan is developed by the MRP method, the upstream workshop produces parts necessary for the downstream process. Figure 6.8 is an example of cylinder assembly lines whose daily production targets remained constant and consisted of eight different cylinders.

To establish production plans and orders it is essential to know how the accumulated production compares with the planned quantities to manage future production. The accumulated production for a given day compared with the resultant accumulated production shows either a shortage or an excess. The percentage achievement is displayed as shown in Fig. 6.9.

```
                    Process : Upper Cylinder Assembly VTR
```

No	Nomenclature P#	08/89 Planned	08/89 Result	08/89 Progress	Order Q'ty	Planned 08/18	08/19	08/20	08/22	08/23
1	AAAAAA 100000	320	320	0	320	320	320	320	320	320
2	BBBBBB 200000	1100	1120	20	1080	1100	1100	1100	1100	1100
3	CCCCCC 300000	255	260	5	250	255	255	255	255	255
4	DDDDDD 400000	150	150	0	150	150	150	150	150	150
5	EEEEEE 500000	510	500	-10	520	510	510	510	510	510
6	FFFFFF 600000	390	385	-5	395	390	390	390	390	390
7	GGGGGG 700000	1150	1155	5	1145	1150	1150	1150	1150	1150
8	HHHHHH 800000	620	600	-20	640	620	620	620	620	620
	Total	4495	4490	—5	4500	4495	4495	4495	4495	4495

```
F9 List    F16 End    ↑↓Scroll
```

Fig. 6.8. Upper cylinder assembly order display.

```
    Process : Upper Cylinder Assembly
```

No.	Nomenclature	P#	Work-in-Process	Rejects
1	Dumper	100000	88	3
2	Cover	200000	340	14
3	Shaft	300000	520	6
4	Disc	400000	22	0
5	Transformer R	500000	330	12
6	Transformer S	600000	125	3
7	Bearing	700000	355	10
8	Motor	800000	751	2

```
    List    [F16] End    [↑↓] Scroll
```

Fig. 6.9. Work-in-process renewal display.

6.6 Conclusions

The system is implemented at the parent factory in Japan, as was noted earlier. It was developed jointly by the Production Engineering Research Laboratory and the Tokai Works of Hitachi Ltd. Although there are quite significant differences between the various video cylinders, the basic technology and processes are identical for most VTRs. As a result, the control systems applied in the Japanese factory may be used in any of Hitachi's other production facilities. However, there are language problems when the production is outside Japan. Owing to the peculiarities of the Japanese software environment which are due to the very different language forms, significant changes have to be made in the realizations used elsewhere.

The strengths of the system are the MRP production control procedure and interactive correction of the plan. These concepts are extensible to cover a large range of products in many categories throughout the world.

7 Anthropocentric Intelligence-Based Manufacturing

T. Ihara

7.1 Introduction

When designing a flexible computer-integrated manufacturing system (FCIMS) which is to be used worldwide, there are many problems that need to be addressed. These may relate to the cultural differences which exist in various regions or countries. Implanting technology designed for use in specific circumstances either ends in failure or friction. It is necessary therefore to understand the different value systems so that they may be built appropriately into manufacturing processes. "Understanding" in this context means clarifying the judgemental processes so that desirable manfacturing goals may be achieved [1]. This understanding may lead to the development of a manufacturing style which is suitable, say, for implementing flexible manufacturing systems in Asia. In essence, then, an aspect of anthropocentric intelligence-based manufacturing is to assist international co-operation in manufacturing by technology transfer. Social, cultural and enterprise factors implicitly influence the form of manufacturing technology used. The industrialized nations are moving towards increasing manufacture of human-sensitivity-oriented products and aesthetic objects. This form of manufacture may be interpreted as an application of anthropocentric intelligence-based manufacturing. Its core technology is the use of computer representation of the knowledge of experienced engineers to control processes.

Human judgement processes are, however, closely related to preferences, psychology and habits. Difficult cases are frequently encountered in manufacturing engineering when specialized methods and systems are encountered. Methods of analysis and synthesis for manufacturing systems which take account of these problems are necessary. There are to date no widely accepted methods for developing anthropocentric intelligence-based manufacturing systems. Some pioneering research has been carried out around the world, which may provide a basic model for this form of system design. This chapter reviews the current state of this research and projects its possible outcome. This chapter also emphasizes the need to represent the knowledge of production engineers in computing systems.

7.2 Concept of Thought Object- and Thought Model-Based Manufacturing [2]

As was mentioned in Chap. 1, there are two current lines of research in anthropocentric intelligence-based manufacturing. Science-oriented manufacturing seeks to judge processes by their end results rather than by their causal relations. The approach found mainly in Europe envisages a "comfortable" environment for both man and machine by developing pleasant surroundings within factories. It pays attention to the interfaces between worker and machine. The other approach, which predominates in Japan, strives to make the machines appear "human" by analysing human elements. For example this may involve analysing the thought patterns of an engineer and how that engineer conducts the decision-making process. Such factors are in the realm of knowledge of an experienced engineer, and may be represented by a "thought object". Manufacturing systems which incorporate such knowledge are therefore described as thought model based.

The thought object is therefore a promising basis for attaining the desired manufacturing goals. Such systems may be expected to be able to manufacture value-added products. Both the thought object- and thought model-based manufacturing have the following characteristics:

1. They apply cognitive psychology, information and knowledge engineering to manufacturing engineering. They make reasonable approximations to an engineer's knowledge of manufacturing and may surpass conventional methods.
2. Object-oriented analysis is used to understand the knowledge and thinking patterns of an experienced manufacturing engineer. Thought objects are derived from thinking patterns, represented as directed graphs, and converted into suitable models. These models do not represent causal relations, rather they are "political-scientific" strategies (i.e. means to an end).
3. As a result of the modelling, a prototype may be defined on a computer as a thought object, although not necessarily in an object-oriented language. This prototype is configured in a manner suitable for application in a manufacturing system.

This method has been applied in certain areas of machine design, process planning, and CAD/CAM interfacing, yielding notable results. The next section discusses its perspective and procedures.

7.3 Core Technologies to Organize Knowledge in Thought Model-Based Manufacturing

7.3.1 Organized Knowledge in Thought Objects

Cognitive psychology asserts [3] that cognition occurs as a result of receiving a stimulus. The stimulus is transmitted from a sense organ and sent to the core. This is a bottom-up view. On the other hand, taking a top-down view, factors such as accumulated experience, motivation, and relative context may significantly influence cognition. Cognition is then the intricate interaction of both bottom-up and top-down processes. Figure 7.1a shows the procedure to fix a bolt. A nut and washer may readily be recognized in this context. However, when the bolt is removed, as in Fig. 7.1b, they become less obvious. For them to be clearly discernible, the detailed illustration in Fig. 7.1c is needed.

Even in such a simple example, the interaction of the top-down and bottom-up processes is evident. Just as the comprehension of a context is dependent on a large body of knowledge acquired empirically, knowledge is crucial to top-down processing. Regarding the extraction and analysis of deep knowledge, Chen and Ito [4] have proposed a notable method. This analyses, extracts and arranges the bodies of knowledge of experienced engineers using a directed graph representation to describe their thinking patterns. These graphs are then converted into models for use in manufacturing. This model consequently does not represent all the information from manufacturing to assembly and production at the bottom as do conventional CAD/CAM systems. Nor does it simply generate necessary information from bottom-end data by the use of algorithms. The information generated at the top is produced with some reference to the bottom-end data. This method avoids the autocratic conventional control used by systems using minicomputers. The required information is produced analogously to the methods used by thinking engineers. Chen and Ito's method was developed for the machine tool industry, which we term an "already mature" industry. By contrast, developing industries, such as mechatronics, have other characteristic aspects. Their problems are those related to products with short lives, frequent introduction of new technology and production of

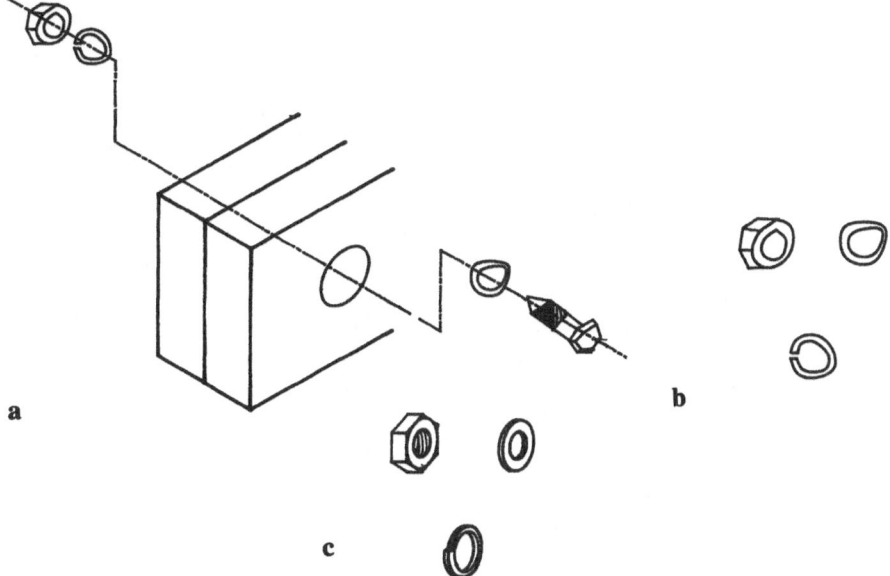

Fig. 7.1. Existence of schema.

maintenance free products. Although these are included in "concurrent engineering" which is used in mechatronics and its related industries, thought model-based engineering has a wider scope.

Cognitive psychology describes the knowledge required for top-down processing as being organized in structures or captured in images for its long term retention (Fig. 7.2). Broadly speaking, structuring knowledge ascribes theoretical significance (meanings) to it. Forming knowledge into images makes it comprehensible as patterns of shapes and positions. The formal structures of organized knowledge are called "schema". Analysis of the thought processes of experienced engineers is the extraction of problem-solving schema (the reasons for reaching conclusions).

The Chen and Ito method expresses a schema as a directed graph whose nodes signify problem elements. This concept of a node is analogous to that of an "object" in object-oriented analysis [5, 6]. It should, however, be called a "thought object", because it may comprise subthoughts for subproblems and contains the concept of function diversification. An example of the representation of the flow of thought of experienced process planners is shown in Fig. 7.3 [4, 7]. Cognition in these diagrams is activated by stimuli. Meanwhile, production of a directed graph is described below to show how the experience and knowledge of an engineer may be extracted. This explanation shows how the information is transformed into the directed graph representation.

Experienced process planners were first given drawings and information as stimuli and then interviewed by means of questionnaires. Directed graphs of the judgement processes were configured on the basis of their responses. Although a degree of vagueness and ambiguity enters the process of knowledge extraction, thought objects may be isolated as elements of the

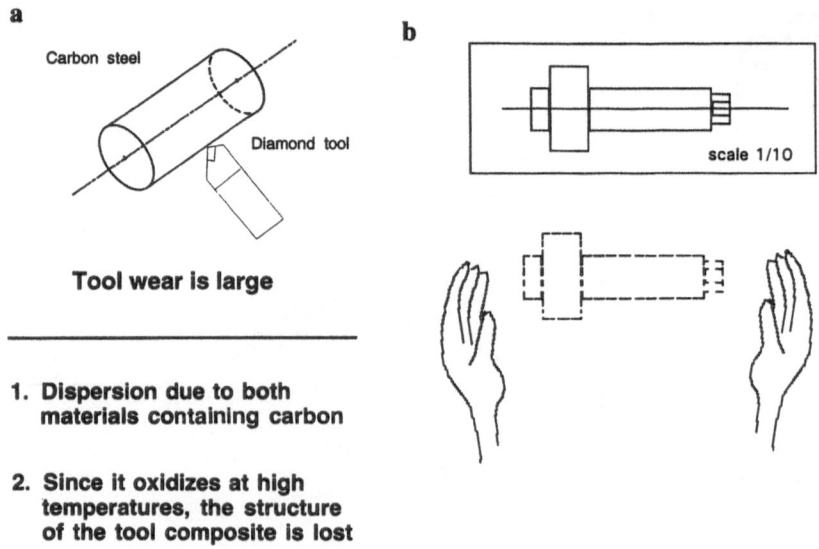

Fig. 7.2. Long term memory. **a** Organizational factor; **b** imagery factor.

process function. Psychology's experimental methods are iteratively applied to reduce the vagueness in the process description. An eye-mark camera may be used to track the motion of the process planner's eyes [2], so that visual motion may be related to the sequence of the planner's examination of the problem in hand.

Thought objects describe process functions. For example, a process function may be: "proposition – the material's dimensions are set to φ 150 × 300" will set the object so that the attribute value φ 150 × 300 is obtained. Other aspects, such as state transitions and procedural knowledge regarding the thought object, and passing messages based on the directed graph, are managed in the normal manner for object-oriented models. In practice, however, it is not necessary to adhere to object-oriented languages strictly because this model is not exclusively used in object-oriented systems. The implementation details depend very much on the computer environment in which they are used, and may be represented in an algorithmic form.

7.3.2   Some Analyses Based on Models of Thought Objects

Although "Means to an end analysis" [8] is a well-known model for schema, real problems require specific knowledge. However, the problems that require cognitive psychology to analyse their specialized knowledge are mostly arithmetical and few in number [9]. The method of using thought objects is believed to be one of the few methods appropriate for the extraction of specialist knowledge from production engineers by the application of cognitive psychology.

The previous section outlined a method proposed by Chen and Ito for investigating the thinking patterns of experienced engineers so that it could be represented in a directed graph form. This may be extended to attempt to solve the most difficult problem in process planning. Even using computer-aided process planning (CAPP), problems relating to complex machining are not evaluated: the problem has proved so far to be intractable.

The graphical algorithm is suitable for static analysis of the directed graph; Petri nets may be used for dynamic analysis. The number of loops may for example be expressed as:

$$\text{(number of loops)} = \text{(number of links)} - \text{(number of nodes)} + 1 \qquad (7.1)$$

Equation (7.1) may be used to evaluate the difficult areas represented in Fig. 7.4, providing that the problems may be represented in a directed graph [10]. As the diagrams demonstrate, it is possible to evaluate the number of loops using the equation: the count provides one indication of the degree of difficulty involved in planning that process. Comparing the results of the loop counts with the experience of a competent engineer, we find:

- The difficulty may be classified as due to either elaborate or highly skilled machining.
- The degree of branching corresponds to the degree of complexity.
- The level of skill required correlates with the number of loops.

The thought model is an effective model, therefore, to analyse the skill of an engineer.

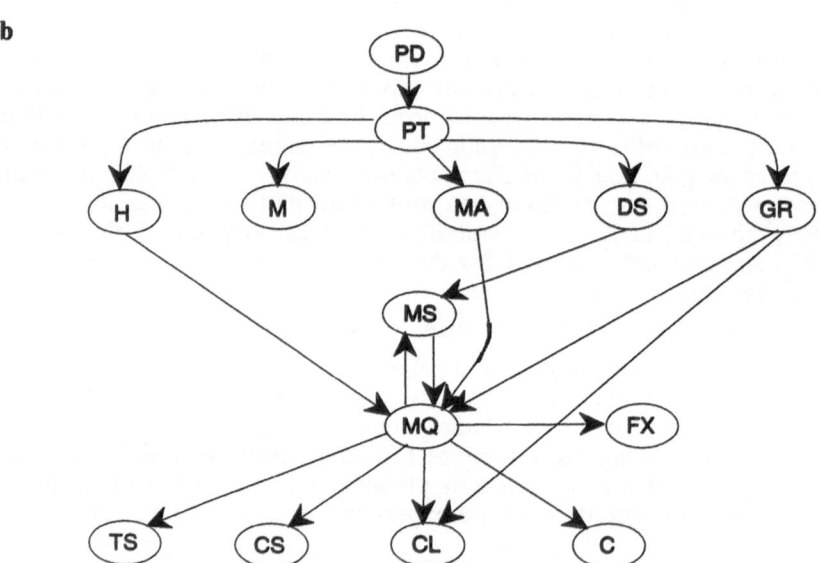

Fig. 7.3. Thinking patterns of experts when providing them with stimulus in the form of part drawing [4, 7]. **a** First type; **b** second type; **c** third type.

c

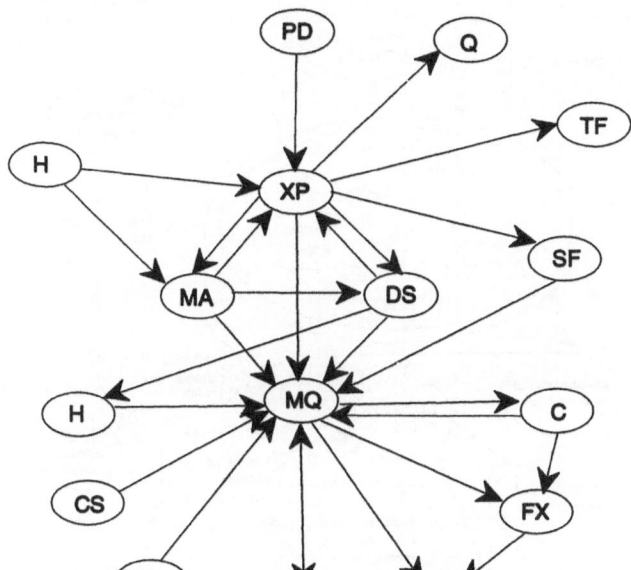

Symbol	Meaning
C	Chucking
CA	Consideration on the re-chucking procedure
CC	Selection of cutting condition
CL	Selection of chucking length
CS	Selection of chucking portion
D	Deadline
DR	Check with experience data
DS	Detailed shape and dimension
FX	Jig and fixture
H	Heat treatment
IL	In-house material list
L	Lot size
MA	Material
MQ	Selection of machining sequence
MS	Selection of raw material size
MT	Selection of machine tool
P	Priority
PD	Part drawing
PN	Penalty
PP	Process planning sheet
Q	Quit the job
RP	Rearrangement of process planning
SF	Selection of factory
TF	Task force
TS	Tool selection
XP	Experience data

Fig. 7.3. (*Continued*)

a

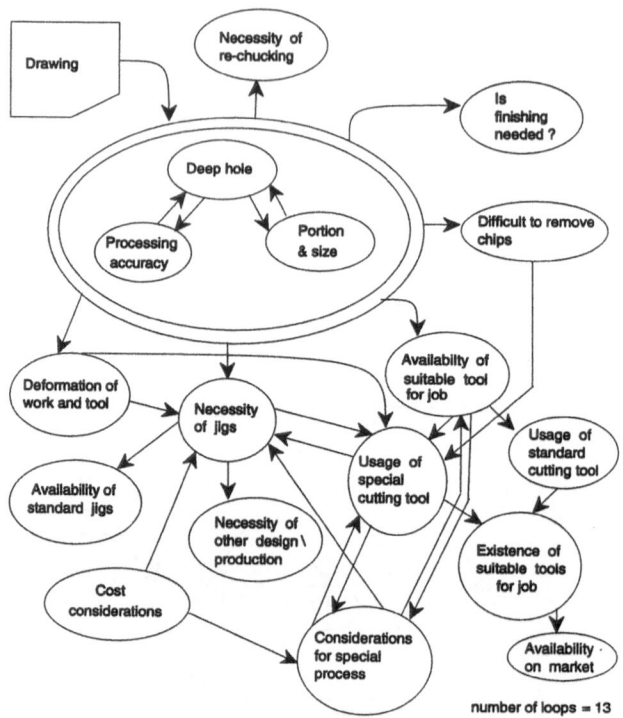

b

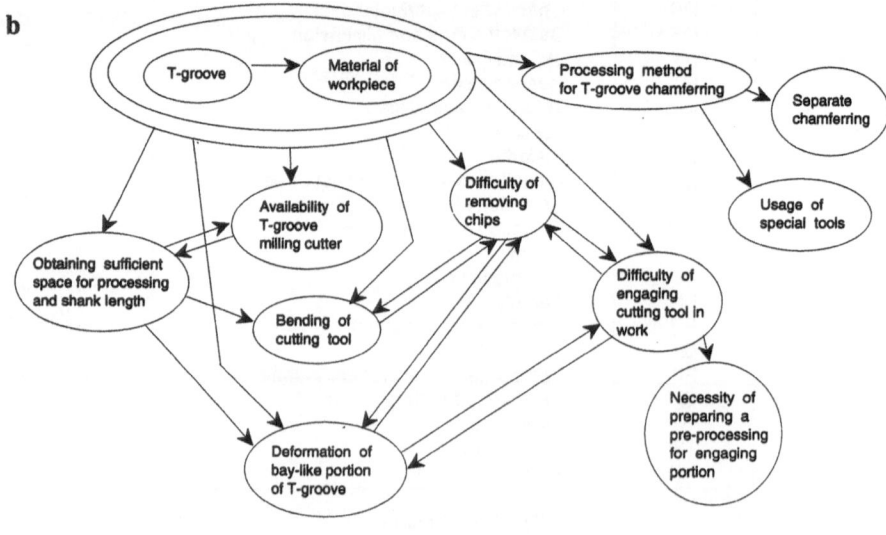

Fig. 7.4. Schema for planning the process. **a** Deep hole; **b** T-groove.

```
***---------------))) PROCESS SHEET (((---------------***

**----------------)) Material Size ((----------------**

Length:   358
Diameter: 60

**--------------)) Machining Sequence ((--------------**

Machining Method No1  LFC (facing)
Machine Tool: 'Turning Center'
Primitive of Machining:   ( 110 )
Clamp Primitive:   Raw Material
                   ( direction: right   position: 144 )

Machining Method No2  DCN (center drilling)
Machine Tool: 'Turning Center'
Primitive of Machining:   ( 120 )
Clamp Primitive:   Raw Material
                   ( direction: right   position: 144 )

Machining Method No3  DS (counter boring)
Machine Tool: 'Turning Center'
Primitive of Machining:   ( 130 )
Clamp Primitive:   Raw Material
                   ( direction: right   position: 144 )

Machining Method No4  L (turning)
Machine Tool: 'Turning Center'
Primitive of Machining:   ( 1 2 3 4 5 6 7 8 9 )
Clamp Primitive:   Raw Material
                   ( direction: right   position: 144 )

Machining Method No5  L (turning)
Machine Tool: 'Turning Center'
Primitive of Machining:   ( 1 )
Clamp Primitive:   Raw Material
                   ( direction: right   position: 144 )

Machining Method No6  L (turning)
Machine Tool: 'Turning Center'
Primitive of Machining:   ( 2 )
Clamp Primitive:   Raw Material
                   ( direction: right   position: 144 )

Machining Method No7  L (turning)
Machine Tool: 'Turning Center'
Primitive of Machining:   ( 3 )
Clamp Primitive:   Raw Material
                   ( direction: right   position: 144 )

Machining Method No8  L (turning)
Machine Tool: 'Turning Center'
Primitive of Machining:   ( 4 )
Clamp Primitive:   Raw Material
                   ( direction: right   position: 144 )
```

Fig. 7.8. (*Continued*)

drawing. Converting the knowledge into images, as shown in Fig. 7.2b is a particularly good example as it does not demonstrate theoretical knowledge, rather it shows the results of episodic memory. The application of this knowledge to thought model objects is therefore problematic. Images are an incarnation of thought different in essence from reality, so their transformation into images representing quantified knowledge is difficult to achieve. The description of the images in abstract language is considered to be the best means to convey their characteristics [13]. Thus knowledge has for now to be extracted by propositional interpretation of the images using directed graphs to represent forms of expression.

As part of the research undertaken into the development of a flair-based CAPP system, experienced engineers were interviewed to find what information they concentrated on when planning machining processes. The results are shown in Fig. 7.9: the encircled elements are knowledge in the form of images. Using this type of knowledge, engineers seem to configure the pictorial information logically. They complete the planning of the process in a short period without processing the intricately detailed portions of the drawing using bottom-up procedures [2].

An eye-mark camera test was carried out to check the validity of the interview results. Figure 7.10 shows the method of the test which studied eye movements of a subject fitted with a pair of "eye-mark glasses". The ocular movements are recorded on videotape and analysed by computer to determine the sequence of movements and duration of views on locations on the drawings. This test may provide information about the viewing

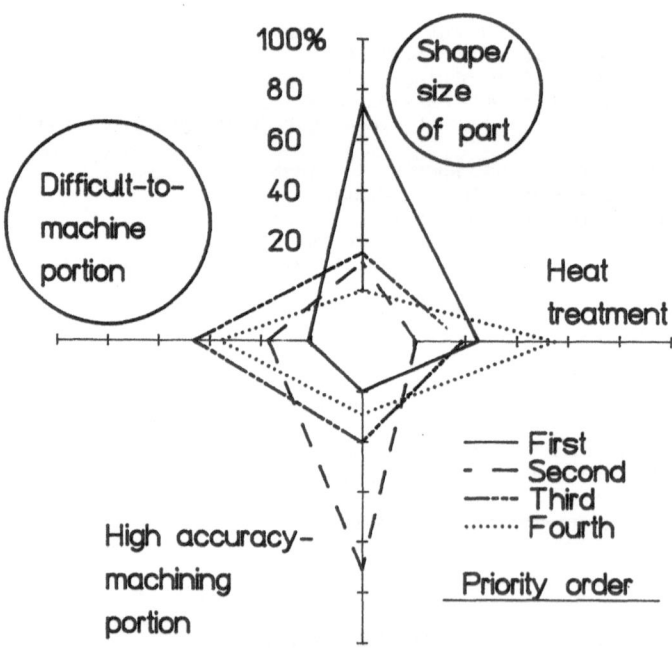

Fig. 7.9. Prioritization on decision-making factors in higher order.

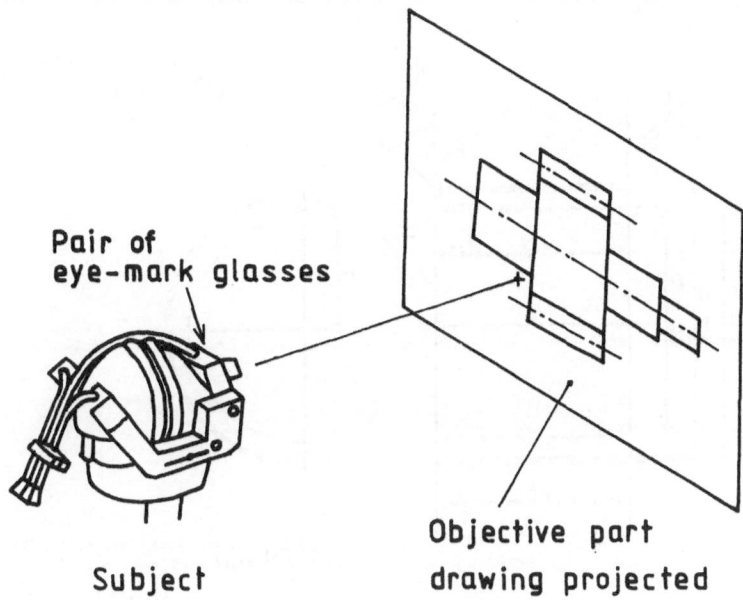

**Pair of
eye-mark glasses**

Subject

**Objective part
drawing projected**

Fig. 7.10. Schematic view of eye-mark camera test.

sequences from both the pinhole and bird's eye viewpoints. In the experiments each engineer was required to produce a process plan whilst observing a part drawing displayed on a screen by a slide projector. Figure 7.11 shows both the drawings and ocular test results for an ordinary engineer. The test information was recorded for 25 s, starting some 10 s after the subject was first shown the drawing.

It should be noted that there is a concentration of focused points which are indicated by a thick black arrow. We conclude that this portion is significantly important in the process planning. However, this region is a key way which has insufficient allowance for tool recessing. Whilst each engineer viewed the circle differently, all paid special attention to the same regions.

Figure 7.12 shows the average time required to produce a process plan. It emphasizes that there is a large difference in the times found for experienced and ordinary engineers. The former required 30 s, only about 1/10 of the time required by the latter group.

Following the tests the priority of decision-making factors involved at the higher levels was also investigated. The results, shown in Fig. 7.9, demonstrate that factors such as the size and shape of the part were of highest priority.

7.4.3   Flair-Based CAPP Systems

As indicated in the last section, engineers perform some tasks quickly. It should be emphasized that these were not simple routines which had been

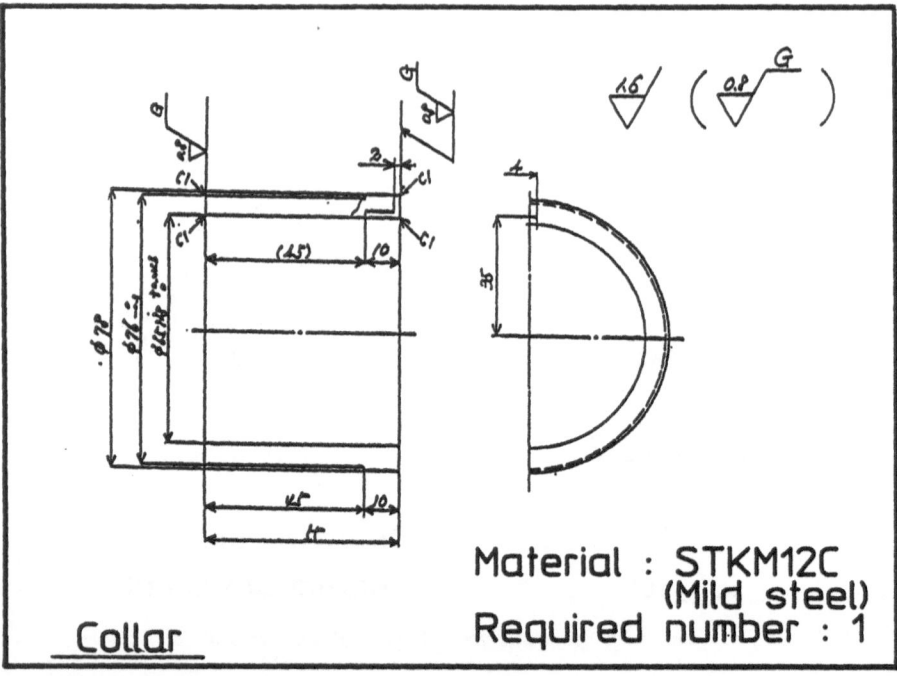

Material : STKM12C
             (Mild steel)
Required number : 1

Collar

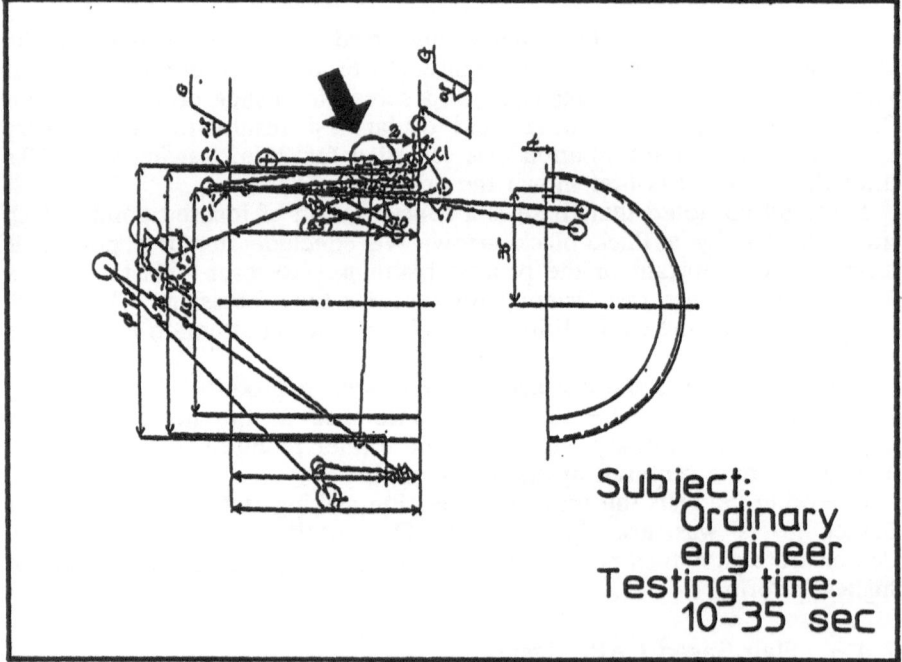

Subject:
      Ordinary
      engineer
Testing time:
      10-35 sec

Fig. 7.11. A result of eye-mark camera test and its objectives.

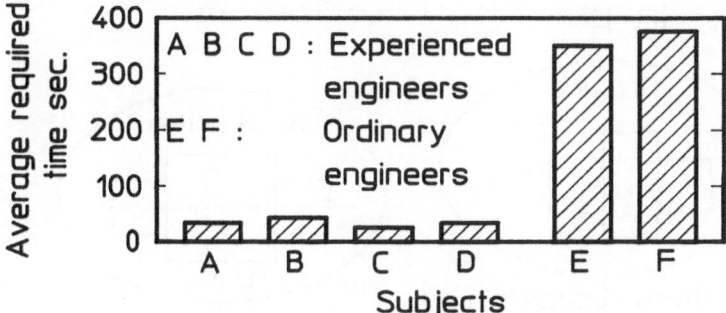

Fig. 7.12. Average required time for process planning.

previously practised [14]. The processing of their decisions, such as the selection of portions of their drawings and deciding which parts required the most attention, demand a high level of intelligence and cannot readily be automated. In a flair-based CAPP system portions requiring special attention are identified and placed into priority order. Further, engineers' knowledge of processing was categorized into priorities. Finally, propositional descriptions of the images were arrived at with the "special attention" portions expressed as thought object models. The orientational links between the objects were generated in the order of their priorities. This description formed the basis of the development of the flair-based CAPP system whose configuration is shown in Fig. 7.13.

In implementing the first version of this system the propositional descriptions of the images were implemented as frame expressions and general order was generated by production rules. Although the problem of propositional interpretation of drawings serving as stimuli remains, this system which uses "model-based inference" is applicable within the manufacturing systems environment. Model-based inference can then produce a process plan for specific drawings using empirical knowledge described in terms of a pattern recognition method.

Figure 7.14 describes the concept of the entire flair based CAPP system. The present prototype routes (3) and (4) are shown in Fig. 7.13, where the difficulty of process planning is relatively low. It will be necessary to prepare the implicit and explicit knowledge bases shown in Fig. 7.14: routes (1) and (2) correspond with the cases of greater difficulty. The design of such knowledge bases is being developed by applying a recently proposed method for quantifying the "difficulty of machining" [10]. We expect that the difficulties of process planning may be evaluated by such a method, in which the number of loops correlates closely with the difficulty of machining as described by Eq. (7.1).

7.5 Conclusion

This chapter has paid particular attention to recent research results for thought objects and models. These, by progressive application of cognitive

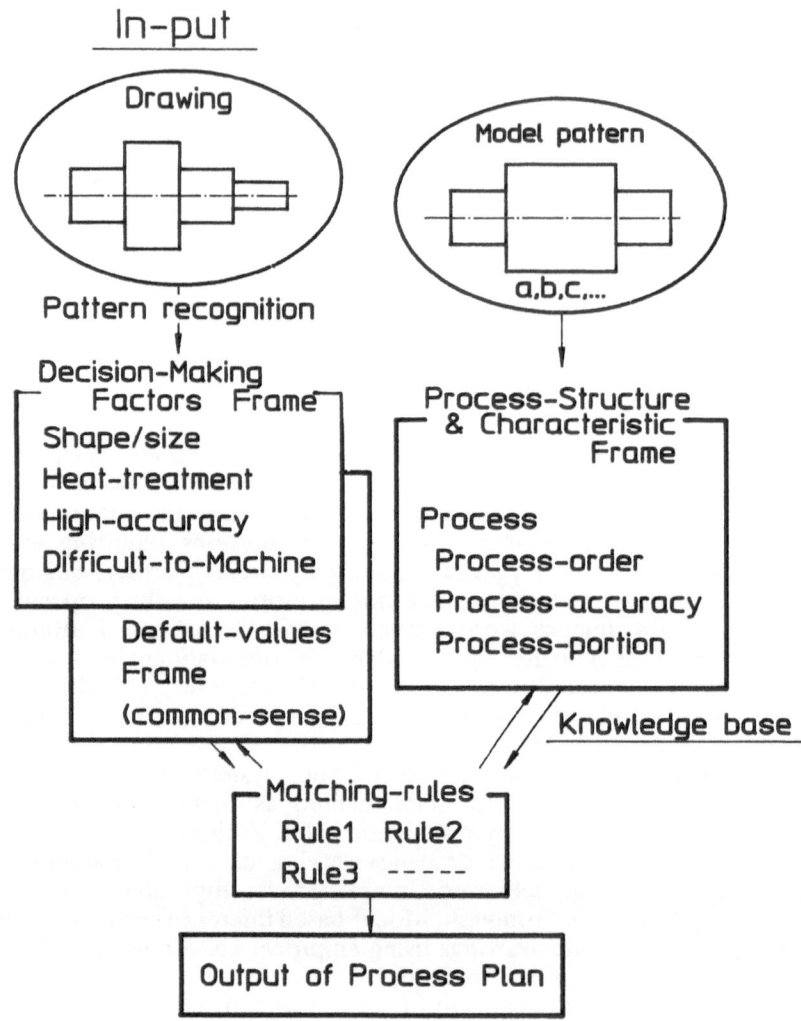

Fig. 7.13. Model for decision-making process of experienced engineers.

psychology, information knowledge technologies and production technology, achieve a level of analysis and synthesis of engineers' knowledge.

This new approach analyses the thought process of engineers using an object oriented model, in which the object is extracted and the model constructed from directed graphs according to strategies used in political science.

The approach holds considerable promise for clarifying the design process so that it may reach the most satisfactory conclusion based on organized knowledge. Although still at the research stage, it shows how knowledge retained in images may be processed. The resulting processes are believed to be an invaluable foundation for future system design and operation in manufacturing.

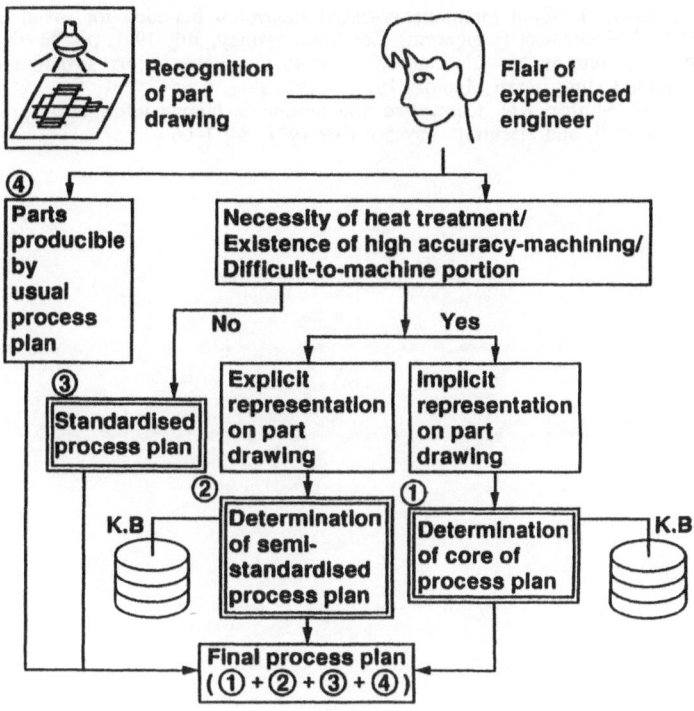

Fig. 7.14. Concept of flair-based CAPP.

References

1. Moritz EF, Ito Y. Computer aided production management – innovation and design. In: 6th international conference on computer-aided production engineering, The Royal Society, London, 1990 pp. 1–8
2. Ihara T, Ito Y. A new concept of CAPP based on flair of experienced engineers. Ann CIRP 1991; 40: 437–440
3. Anderson JR. Cognitive psychology and its implications: WH Freeman, San Francisco London, 1980
4. Chen MF, Ito Y. Investigation on the engineer's thinking flow in the process planning of machine tool manufacturer. In: 13th NAMRC, UC Berkeley, 1985; 418–422
5. Coad P, Yourdon E. Object-oriented analysis, 2nd edn. Prentice-Hall, Englewood Cliffs, NJ, 1991
6. Shlaer S, Mellor S. Object-oriented systems analysis. Prentice-Hall, Englewood Cliffs, NJ, 1988
7. Chen MF, Ito Y. A proposal of data base structure for automatic process planning with expert system. In: 14th NAMRC, University of Minnesota, 1986, pp 523–527
8. Newell A, Simon HA. GPS, a program that simulates human thought, In: Feigenbaum EA, Feldman J (eds) Computers and Thought. McGraw-Hill, 1963
9. Kintsch W. Understanding word problems, linguistic factors in problem solving. In: Nagao M (ed), Language and artificial intelligence. North-Holland, 1987
10. Ihara T, Ogawa M, Ito Y. An evaluation method for difficult-to-machine portions. In: Proc. JSPE, Tokyo, 1992
11. Ito Y, Shinno H, Nakanishi S. Designer's thinking pattern in the basic layout design procedure of machine tools. Ann CIRP 1989; 38: 141–144

12. Ihara T, Sando T. Goal maintenance-based operation planning for metal cutting. In. International Mechanical Engineering Congress, Sydney, July 1991, pp. 93–97
13. Mandler JM, Johnson NS. Remembrance of things parsed: story structure and recall. Cognit Psychol Hum Learn Memory 1977; 3: 386–396
14. Schneider W, Shiffrin RN. Controlled and automatic human information processing: 1, detection, search, and attention. Psychol Rev 1977; 84: 1–66

8 Culture and Success in Manufacturing: a Comparison of Japan and the USA

M.J. Kolar

8.1 Cultures are Formed to Alleviate Human Anxiety

It has been argued that a nation cannot have a strong economy without a significant contribution from the manufacturing sector [1]. In the past 30 years, the world has seen major shifts in the manufacturing strengths of nations. For example, the USA has suffered a long decline in the percentage of its GNP produced from manufacturing while Japan has seen a marked increase. There are many factors which influence these trends, such as savings rate in a country, value of the local currency relative to other currencies, the availability of workers, training, management, and so forth. There have been many articles and books written examining these factors (e.g. [2]) and, since the mid 1980s, an increasing number of books have appeared on the role of culture in corporations (e.g. [3, 4]. More recently, books have begun to appear on the effect of national, regional, or local culture, on a country's chances of being successful in manufacturing (e.g. [5]).

At first glance, culture may seem to be a relatively unimportant factor compared to, for example, financing. In developed countries, the number of people involved directly in production has been decreasing for many years as evidenced by the relentless increase in "productivity" [6]. Consequently, one might think that cultural issues, which primarily affect human beings, would be of relatively little importance. This is not the case for two very important reasons. First, irrespective of how many people are involved in direct production, humans will always play a key role. Second, manufacturing involves far more than direct production, it is a people-intensive enterprise.

To expand a bit on these two points, consider the following. Production cannot be "fully automated". The idea that this is possible arose primarily because of the birth of the fields of "artificial intelligence" and "robotics". Many manufacturing managers, particularly in the USA, bought into these technologies, apparently without fully comprehending potential limitations and consequences. For example, General Motors invested billions of dollars in factory automation only to find that, after the process was redesigned so that machines could do the job, humans could produce the product using the same process more efficiently than the machines [7]. There are two underlying causes that prevent machines from completely replacing humans in production; flexibility and capital cost. For example, robots show no sign of even approaching the flexibility of humans. In the case of artificial intelligence, the initial cost of developing an expert system is far too high to produce a pay back in a realistic time for most manufacturing applications.

Production, however, is only a small part of the manufacturing process in the USA.[1] Within the company, finance, sales, marketing, product development, design, production, and shipping are all involved. Outside the company, suppliers, customers, competitors, national infrastructure, national government, foreign governments, and financial markets are involved. One simply cannot ignore the role of people in manufacturing and, since culture is crucial to the functioning of humans, culture must be understood and utilized properly by any successful manufacturing enterprise.

What is culture? Why do humans develop cultures? What are the key elements of culture? How can a manufacturing enterprise take advantage of culture to gain an edge on the competition? These questions, and their answers, are at the centre of this chapter.

The dictionary [8] defines culture as "the integrated pattern of human knowledge, belief, and behavior that depends upon man's capacity for learning and transmitting knowledge to succeeding generations . . . the customary beliefs, social forms, and material traits of a racial, religious, or social group." So culture involves beliefs and behaviour. How does this translate itself into an industrial setting? An example from the author's experience shows how differences in culture can impact upon a business enterprise [9].

At approximately 4:00 a.m. on a March morning in 1979, an unlikely sequence of events began to unfold at a power station near Harrisburg, Pennsylvania. While operators debated how to resolve a problem that had occurred, a power plant self-destructed leading to, among other things, a financial disaster for the plant's owner. The aftermath of this debacle led to many technical, administrative, and regulatory changes in the entire power industry. Yet little, if anything, was said in public about one of the most innocuous underlying causes of the Three Mile Island accident – a clash of cultures within the organization.

There are two reactors at the Three Mile Island site. Unit 1 was initially staffed by a cadre of people drawn primarily from other power plants. These individuals had a common background and culture which, in part, permitted an individual to disagree with a supervisor with the expectation that his ideas would receive serious consideration. The operations of Unit 1 were relatively harmonious.

Unit 2 was staffed by a mixture of people including some that had worked in the power industry, and some who had learned their profession while serving in the military. The latter group included the two operators who were debating how to resolve the problem on that fateful morning. They had left the military hoping to find a less autocratic management structure. In the course of events, their supervisor at Three Mile Island was a former

---

[1] In Europe, the reverse is the case: production is greater than manufacturing.

military officer whose management style was similar to the one common in any military organization. Predictably, relations between the supervisor and operators were somewhat strained and, consequently, the operators were reluctant to place a telephone call to their supervisor – an individual who, in all likelihood, could have suggested a timely course of action that would have prevented the accident.

This rather dramatic example of clashes of culture within organizations is repeated daily, albeit on a more modest scale, in industries throughout the world. An individual's view of how he or she wishes to live and work differs from those of a superior, conflict ensues, with resulting difficulties for the organization. Much has been written about how to deal with individual problems of this sort; but how does a manager establish a milieu in which such problems are less likely to occur? How does one set up a milieu when there are individuals in the organization with different nationalities and values? How does an organization operate when the industry itself is both national and multinational? The approach suggested in this chapter is to understand and use culture – the personal culture of the individual workers, the culture of the organization, the culture of the industry, and the cultures of the countries in which the organization does business. This chapter is directed toward the manufacturing industry, but most of it is equally applicable to other industries.

A common feature of the modern international organization is that it contains people that represent many different national and personal cultures. A manager who attempts to direct such a diversity of people is often limited by his or her understanding of individual cultures. It would be extremely useful to have a common set of underlying principles which would apply to all individuals from all cultures so that a manager could act in a coherent manner, confident that a single message was being heard by all workers. This seems a bit much to hope for but, if a manager could understand what culture accomplishes and how it forms, he or she would have a valuable tool. The main purpose of this chapter is to develop such an understanding and to show how it can be used to understand the culture in manufacturing industries.

In Chap. 7 of his book, *Organizational Culture and Leadership* [10], Edgar Schein gives an excellent review of theoretical perspectives on how cultures form in human enterprise. A common theme involves anxiety. My own experience, which includes management assignments in industry, non-profit organizations, and academia, suggests that understanding and controlling human anxiety is a key – perhaps the most significant key – to successful management. Further, in my view, culture develops specifically to help humans control anxiety. My ideas have been moulded by many years of study and practice and have been influenced by many people; the result has been the beginnings of a theory of management that has been extremely successful in getting people to work together. The elements of the theory were derived as follows.

In 1973, Ernest Becker published a book that was destined to win many awards, including the Pulitzer Prize. The book, *The Denial of Death* [11], was written while Becker was dying of cancer. Building on the work of many philosophers and psychologists, Becker argues that the leading cause of mental illness in individuals is the denial of death. He claims that an underlying existential anxiety is present in all humans that is caused by the fear of death. If this anxiety is not handled properly, then aberrations in behaviour occur. For example; the individual's need to be a "hero" is a manifestation of this anxiety. While there has been a great deal written on death by philosophers throughout the ages, Becker's contribution was to connect the existential anxiety caused by the fear of death and mental illness in individuals.

It is important to understand the meaning of "existential" as used in the context of human beings. An existential feature is something that occurs simply because humans exist. For example, humans occupy space or they think simply because they exist. There is nothing that anyone, including a manager, can do to change this. Consequently, if all humans suffer from an existential anxiety caused by the fear of death, then there is nothing that anyone can do to change that. One could look for ways to reduce or control that anxiety, but one could never eliminate it.

In 1975, a second book written by Ernest Becker, *Escape from Evil* [12], was published, posthumously. In that book, Becker argues persuasively that civilizations become – employing a term (not used by Becker) that is in common use today – "dysfunctional" (i.e., having impaired or abnormal functioning [8]), when the humans who individually make up the civilization become dysfunctional by denying death. One result is war, where two civilizations each believe that they can survive even though, inevitably, one or the other or sometimes both of the combating entities are devastated. He argues that heroes are necessary to help such civilizations survive.

Based on the above, one can postulate that, just like civilizations, organizations become dysfunctional when the individuals who make up the organization are dysfunctional. This idea is not profound and others have suggested it [13, 14]. However, if Becker is correct, then all humans suffer from the existential anxiety caused by the fear of death and, therefore, organizations must take this into account if they are to function properly. Further, when death anxiety is not being handled well, then it is necessary to have heroes within an organization. The same idea extends to industries, nations, and the world community.

From the discussion thus far, we can conclude that all organizations are in danger of being dysfunctional simply because they are composed of humans. In 1980, a book was published that was written by Irvin Yalom, a psychiatrist at Stanford University. The book, which is very readable despite its scholarly format and imposing title, is entitled *Existential Psychotherapy* [15]. In my opinion, Yalom's book should be mandatory reading for anyone who wishes to engage in the art of management. Yalom extends Becker's ideas and gives them a new validity; that of a practising therapist based on actual experiences with human beings. Yalom focused his work on the individual. He confirms that one of the underlying causes of mental illness is the existential fear of death. Yalom goes on however, to list three additional existential fears that plague all humans; the fear of freedom, the fear of isolation, and the fear of meaninglessness.

Yalom explains the difference between the Freudian and the Existential approach to understanding human behaviour as follows:

Freudian

Drive $\longrightarrow$ Anxiety $\longrightarrow$ Defence mechanism

Existential

Awareness of
ultimate concern $\longrightarrow$ Anxiety $\longrightarrow$ Defence mechanism

Kierkegaard defined fear as the fear of *some* thing and anxiety as the fear of *no* thing. To avoid anxiety, we replace it with fear, i.e., we invent things to worry about. (To see this in yourself, consider how the importance of business matters diminishes in one's mind after a vigorous physical workout or a long weekend in the country.)

Yalom expands significantly on how the awareness of ultimate concerns manifests itself in individuals. The fear of death plays a major role in our internal experience. The child, at an early age, is pervasively preoccupied with death. We erect defences against death awareness, based on denial, that shape character structure and, if maladaptive, the result is mental illness.

Yalom suggests that one common way that the fear of death manifests itself is that each of us, first as a child and then as an adult, clings to the irrational belief in our own "specialness". Compulsive heroism, the workaholic, narcissism, and aggression and control are all manifestations of specialness. He offers numerous examples of dramatic changes which occur in people when they are suddenly confronted with the reality of their own mortality.

Another way that the fear of death is dealt with by individuals is belief in what Yalom calls an "ultimate rescuer". We all want to believe that someone will take care of us, e.g., the parent, the mentor, the boss. In the mind of the individual, this entity is endowed with powers that will provide protection from death. Yalom concludes his discussion on death-related anxiety by relating the inadequate response on the part of an individual to the existential fear of death to many forms of mental disturbance.

Freedom, according to Yalom, is another ultimate concern of humans which produces anxiety. Stated briefly, we each truly are responsible for deciding how we will live. Responsibility is the operational word here. We create our own destiny, life predicaments, feelings, and sufferings. If you insist on blaming other individuals, other forces, or your own "disposition", you cannot truly live.

Yalom emphasizes that we are responsible for our own happiness. When we do not live as fully as we are capable, we experience existential guilt. Most managers will be familiar with Maslow's idea that the highest level of achievement is "self-actualization". However, few are aware that when a human does not reach this level of achievement, (s)he suffers existential guilt which gives rise to anxiety.This anxiety makes it difficult for humans to reach decisions. The underlying fear that the choice may not be optimal, that we must cut off the paths that we do not choose, immobilizes people. Decisions shift the ground underneath us – or show that there was no ground there at all. Making a decision is a lonely act; no one else can decide for us. The resulting anxiety can only be overcome by an act of will.

Yalom makes the point that response and ability are both essential to responsibility. He quotes a Japanese proverb: "To know and not to act is not to know at all". A vast gap exists between willing to do something and wishing or wanting to do it. Inability to act causes us to lose our ability to feel emotions, we lose spontaneity, and become wooden, heavy, lifeless, boring. A common response to decisions is to avoid renouncing our choices. We do not let go of everything. (As, for example, when a person is promoted from a technical to a managerial position but insists on "keeping their hand in the technical end".) We devalue the alternatives. (As, for example, when we buy a new car and become convinced that all other choices would have been completely inferior.) Or, most common of all, we delegate the decision to someone else, pretending that we, therefore, have not made a decision.

Yalom discusses the history and background of the concept of freedom and gives numerous clinical examples to help see how the anxiety caused by this existential reality displays itself in human actions. The fear of making decisions has many implications in organizations. It can cause excessive delays in selecting a final design or in choosing a marketing strategy. It can also keep people from leaving organizations even though they are unhappy.

The third ultimate concern put forth by Yalom is "existential isolation", the fact that there is a gulf that cannot be bridged between oneself and any other being. The connection between death and isolation is that we each come into this world alone and we will each go out of it alone. Freedom and isolation are related in that we alone are responsible for the choices we make in life. We are, in an existential sense, our own parents. There is nothing that we can do about this essential isolation from all other humans. This is a very lonely position which all humans must face. If we do not face it squarely, anxiety follows. The desire for humans to belong to a group or to be valued by others is a manifestation of this existential anxiety.

The fourth, and final, ultimate concern addressed by Yalom is meaninglessness. Humans seem to require meaning, yet freedom implies we are choosing our own lives. Cosmic meaning is sought through religion or the search for a Creator. The human search for meaning often becomes a personal, secular search for meaning. Instead of answering the question "Why do we live?", we answer the question "How do we live?". We establish causes – the state, our family, our company, our friends, etc. This sense of meaninglessness gives rise to an existential anxiety within us. Creativity is a powerful antidote to this anxiety.

8.2 National Culture Affects the Likelihood of Success in Manufacturing

The central thesis of this chapter is that cultures arise to enable humans to deal with the four existential anxieties suggested by Yalom. Mature cultures are those that succeed in dealing with these anxieties in a constructive manner. Manufacturing industries in countries with mature cultures are in a better position to succeed than those in countries where the culture has not matured. However, both the industry and the organizations within it must take advantage of the culture.

We may group countries into those with a single predominant culture and those with multiple cultures. Countries in the first category are those such as Japan, Germany, Mexico, etc., where the vast majority of people have a common ethnic origin. Countries in the second category would include the USA and Canada, where people from many or several predominant ethnic origins live. We will term the first category "single culture" countries and the second category "multiple culture" countries.

Even in single culture countries, dramatic variations occur in culture as one moves from region to region. For example, the culture of people living near or in large cities may differ markedly from that of people living in small towns. Nonetheless, a manufacturing organization located in such a country need only deal with a single predominant national culture. In multiple culture countries, manufacturing companies must deal with both multiple national cultures and regional variations.

Mature national cultures take many, many centuries to develop. At this point, the reader must take this statement on faith but, as we will see later, there are several concrete examples that support this thesis and – to the author's knowledge – no counterexamples.

A culture is created by humans to help them deal with the existential anxieties caused by fear of death, fear of freedom, fear of isolation, and meaninglessness. One can alleviate fear by dealing with it either directly or indirectly. The most direct method – eliminating the source of the fear – is not possible in the case of an existential cause. Educating people to this fact is a good starting point. Most cultures deal with existential anxieties indirectly. For example, myths and rituals develop around death, people are trained to accept the decisions of law or other authority, patriotism or racism is instilled in people, or organized religions are established.

It is the author's opinion that the best way to deal with existential anxiety is directly. In societies where the population is in direct contact with its livelihood, such as agrarian societies or those that derive their living from the sea, there is ample opportunity for this. However, in industrial societies, learning to deal directly with existential anxiety presupposes a level of education which is not available to the vast majority of people. Consequently, indirect methods must be employed. The result is that, in the case of industrial societies, there probably are no completely successful cultures – rather, there are cultures of varying degree of maturity. It is important that a manufacturing organization understands the level of maturity of the nations in which it does business.

In the following sections, the cultures of two countries, Japan and the United States, will be investigated with the intention of determining how they cope with the four sources of existential anxiety.

National cultures are usually described in terms of family, social customs, education, religion, the arts, politics, etc. Indeed, all aspects of life, taken together, could be construed as what we mean by national culture. However, we will take the view that certain aspects of living have greater impact than others on the way the overall culture deals with existential anxiety. In particular, family life, social customs, and education will be viewed as the most influential in building culture, while the arts, politics, etc., will be viewed as manifestations of the underlying culture. With this approach, it is possible to include manufacturing and its organization as a manifestation of the underlying culture.

8.3 Japanese Culture is Relatively Mature

Much of the material for Sects 8.3 and 8.5 was gathered by the author during a six-week study of the culture surrounding manufacturing in Japan conducted in the summer of 1990 [16]. It is a summary of observations which will be integrated in a later section of this chapter. A great deal has been written about the family and social structure of Japan [17, 18]. We list here those factors related to family and social structure which we believe have significant impact on manufacturing in Japan.

The concept of an inner and outer world is learned at a very early age. For example, the Japanese family considers its home to be its inner world and everything outside the house is the outer world. Another example, the Japanese language uses different words to describe a person depending on whether or not that person is a part of one's social group.

A hierarchy according to age is imbued in Japanese children. For example, the Japanese language has different words for brother depending on whether the brother is older or younger than the sibling who is speaking.

Practically all Japanese children have an older family member (usually the mother) who pushes the child to learn and to succeed in school.

The Confucian value of devotion to one's family is common throughout Japan. This leads to the desire of children to succeed in school so that their family will not be shamed (lose face). This value is so deeply embedded, that it is transferred to other groups to which one belongs including one's school and one's company.

Reverence for the dead is learned early and is practised throughout one's life. For example, ancestors are revered in every home throughout Japan. One of the largest National Holidays, Bon, finds all Japanese returning to their home of origin to pay homage to their ancestors. Another example, stone dolls, clothed in baby clothes, are found in many temples throughout Japan; the dolls commemorate aborted children.

The importance of being Japanese is instilled in children at an early age. Because there are very few non-Japanese for children to come into contact with, this concept is reinforced by experience.

Many Japanese are still relatively close to their source of livelihood. Even in large cities like Tokyo, one can find rice growing in the occasional open lot. Further, many Japanese still have relatives who live in rural areas and fishing is a common hobby in addition to being a major industry.

The Japanese educational system is described in Ref. 17. Briefly, grade, middle, and high school education is rigorous, demanding, and broad. The primary and secondary systems come under the purview of the national government so that all Japanese children must meet the same standards. Many Japanese children attend two schools at the same time – the normal full time public school, plus a private school three evenings a week. The competition which students face is fierce. All high school students are trained in mathematics and the sciences as well as in foreign languages (English being one) and social studies. At least 95% of all Japanese children graduate from high school. Japanese high school graduates appear to have studied more science and mathematics than most US college graduates. To enter a Japanese university, a student must pass an extremely competitive examination. The following applies to those students who pass the exam and choose to study engineering.

There are approximately 180 engineering schools in Japan. A handful of these – those that were once the Imperial universities plus a few private schools – are considered to be the elite. Entry into these schools has traditionally meant a secure job for life; engineering graduates of the elite schools tend to enter government, banking, and manufacturing. The emphasis by students at these schools is toward making personal friendships and developing leadership skills rather than studying. Apparently, many foreign observers visit only this group of schools, because much of what has been written about Japanese engineering education appears to be a description of them.

We visited five universities from the group described above, and two from the remaining 170 or so schools. We also spoke to professors from six additional schools in the latter group. Most of the mechanical engineering students from the latter group of schools go into manufacturing. We received the impression that course work, laboratories, and the study habits of students at these schools were comparable to those of US engineering students. Moreover, we were told that most students from this group of schools stayed on for one additional year of study, during which time they obtained a Master's degree.

There are two ways to earn a Doctorate in Engineering in Japan; the common US method of continuing to do course work and thesis, and what is referred to as a "paper" Ph.D. whereby a person with extensive industrial experience who has published many papers submits his/her work to a university as evidence of qualification for the degree.

Several interesting facts arose during our discussions at universities. First, professors felt they have some influence over which company a student chooses to join, but little influence over the student's choice of industry. Particularly at the elite schools in the Tokyo area, banks recruited an increasing number of engineering students over the five years from 1981 to 1986. However, following the October 1987 US stock market crash, this trend reversed itself. Further, engineering graduates who are already working in the banking industry have begun to approach professors to ask for assistance in changing to the manufacturing sector. The apparent reasons are a lack of job satisfaction and concerns over job security. Second, nearly all engineering graduates aspire to work for large companies; this desire is apparently fostered by parents who are concerned about their children's long term job security. A result is that virtually no engineering graduates are interested in starting new companies.

All of the schools we visited reported difficulty in attracting students to curricula aimed at production machinery. At one of the schools we visited, the agricultural engineering department was training students for this type of work. Some schools have solved the problem by restricting other options, effectively requiring some students to choose this curriculum if they want an engineering degree.

There is only one Japanese business school (Keio University, Tokyo) in the entire country. Very few Japanese engineering graduates get a Master's Degree in Business Administration (MBA). As was noted above, a large number of engineering graduates get Master's degrees in their discipline. Further, Japanese manufacturing companies often send engineers overseas to acquire advanced degrees; apparently, only an extremely small percentage get an MBA (e.g., one 35 000 person company we visited had only two MBAs on its entire staff).

Nearly every top manager in the Japanese manufacturing companies which we visited had at least an undergraduate degree in engineering. All had been provided the necessary business-related training by their company.

Japanese culture can be thought of as dealing with the anxiety caused by the fear of death by giving people the assurance that they will not be forgotten after they die. Teaching the concept of filial piety also is a way of instilling the habit of obedience to authority, which reduces the number of decisions that one must make in life.The concept of an inner and outer world gives a child the sense of belonging to a group. That this characteristic of the Japanese is ubiquitous, can easily be seen by observing the Japanese preference for travelling in groups or working in teams. Finally, meaning is provided by instilling the importance of preserving the Japanese race. Clearly, Japanese culture is very old and has developed some very strong antidotes to the four existential anxieties identified by Yalom.

The Japanese education system builds on and reinforces these concepts so that, by the time a Japanese enters the manufacturing industry, he or she has a very solid set of defences built in to deal with the underlying existential anxieties.

8.4 American National Culture is in Very Early Development; the United States has Many Mature Subcultures

Unlike Japan, there is relatively little written about the national culture and subcultures found throughout the USA. There *has* been a great deal written about the American consumer. Marketing experts spend enormous amounts of time and money examining the buying habits of every segment of American society. There also have been some recent books that discuss certain aspects of American culture such as "pop" culture. However, these do not address the entire spectrum of culture in the United States. A recent book by George and Louise Spindler, entitled *The American Cultural Dialogue and Its Transmission* [19], provides a summary of American culture from an anthropological point of view. Such information is of little use to the company trying to locate a manufacturing plant, and there appear to be no books which consider American culture from the point of view taken in this paper. Consequently, a summary from the author's experience may be useful in its own right.

The following information is based on the author's experience of living and working throughout the USA for more than 50 years. It includes long term living in five different states located in the midwest, the east, the far west, and the Rocky Mountain region. It also includes raising five children in those places. In addition, it includes short term work assignments in 35 additional states, technical positions in government, and management positions in industry and academia. It has included living in large cities, suburbs, and on a farm. It also includes a term as a school board president.

In terms of national cultural development, the USA is still in its' early childhood compared with a country like Japan. Furthermore, it has built-in mechanisms that prevent it from developing very rapidly. As is well known, the United States is a country of immigrants and their (relatively recent) descendants. There has long been a dominance of political and economic power by white males of northern European heritage who are of Protestant persuasion. However, there has never been an attempt to develop a national culture that could deal effectively with existential fears. The only, rather weak, approach to this problem has been the encouragement by the government for people to participate in organized religion.

It is almost as if the USA has embarked on a path in which it tries very hard *not* to deal with the problem of existential anxiety. For example, the cult of youth is worshipped. Rather than dealing with the fact that they must die, Americans succumb to advertising campaigns that lead to eternal dieting, hair transplants, a medical industry that consumes 12% of the gross national product, and inner cities where the old who cannot afford to move are preyed upon by an uncaring youth.

Americans are raised to challenge authority. When asked to list their most cherished value, Americans answer "freedom of . . ." where the blank includes speech, mobility, etc. The American legal system leads to a continuing barrage of lawsuits, drugs are a widespread problem despite their illegality, and teachers who discipline unruly students are at risk of being sued. By promoting an "anything goes" image of itself, the USA encourages its youth to make decisions on increasingly meaningless topics. What colour of eye shadow or what kind of basketball shoe to choose are questions that provoke anxiety for little gain. This is not freedom in the sense discussed by Yalom. Americans want freedom but have disdain for the concomitant responsibility. (Look at the low percentage of Americans who participate in elections!)

The American ideal is the rugged individualist, but not the person who accepts his or her existential isolation. The rugged individualist deals with his or her anxiety by stepping on the rights of other people. Dealing with existential isolation requires autonomy – not individualism.

Finally, the USA has tried to deal with the question of meaning by identifying (or inventing if necessary) external "enemies". With the end of the cold war, it is becoming increasingly obvious that the USA has little to offer in the deeper sense of meaning.

How then, could a country whose culture is so immature have been so successful in manufacturing? From the late 1880s until the 1970s, the USA was the most powerful nation in the world in achieving growth in manufacturing. This was achieved by "using up" cultures from people who had been imported.

Immigrants who come to the USA bring with them the culture of their country of origin. During its rise in manufacturing prowess, the major industrial cities of the USA – Detroit, Pittsburgh, Chicago, Cleveland, etc. – were all composed of many neighbourhoods each composed of an ethnic group from rural settings in Europe or from the southern USA. These people had their own ethnic heritage but a shared common culture – that of an agrarian society. The children of these people were one step removed from this common culture and their children's children even further removed. As succeeding generations lost contact with their rural roots, they also lost the beneficial effects that the corresponding culture had in assuaging the four existential anxieties.

As the newest generations matured physically, they began to make demands on business that were unreasonable. Why would relatively well-educated people demand such unreasonable compensation that their own jobs would be lost? Were they seeking more and more material rewards to keep their minds off of the existential anxieties that were gnawing at them?

The American family system is very weak today. While there are a few notable exceptions, the vast majority of American children learn their values from television programmes. Very few children have a full time adult from whom they can learn values. The US primary and secondary schools are controlled by literally thousands of local school districts so there are no national standards on what American children learn. Indeed, local values are often incorporated into the curriculum so that children from different regions learn different values.

American elementary and secondary education produces relatively poor results. About 75% of students complete high school and, of those that finish, only a small percentage have the requisite courses in mathematics and science to allow them to enter mathematics, science, or engineering programmes at universities.

American universities allow students to enter based on either a national exam (the Scholastic Aptitude Test or SAT is the most common), or on class standing in high school. US university programmes in engineering can seek accreditation from the Accrediting Board for Engineering and Technology (ABET), a national board established by the professional societies. Students who graduate from an ABET accredited programme all have at least a specified amount of course work in mathematics, science, engineering, social science, and the humanities.

American society today has relatively few ethnic groups that are still in direct contact with their rural roots. Aside from the influx of Vietnamese in the mid 1970s, there have been no recent mass immigrations. (It is interesting to note that one of the best performing student groups during the last 10 years has been Vietnamese.) Thus America is now composed of third, fourth, and higher still generations who have neither the rural culture of their ancestors nor a national culture that will help them deal with the four existential anxieties.

8.5   Japanese Manufacturers have been Very Successful in Using National Culture

The Japanese industrial system is described in Refs 20, 21, and 22. In summary, the manufacturing industry is commonly composed of large companies that perform design, marketing, sales, and final assembly of products, with much smaller, independently owned, second and third tier companies manufacturing subassemblies and components. Manufacturing employment is split about 25%, 25%, 50% between the first, second, and third tier companies. The second and third tier companies are often completely dependent on a single first tier company for their business. Salaries in second and third tier companies are perhaps 20% lower than those of first tier companies. Both first and second tier companies promise to try to keep their workers employed until retirement (usually age 55), but third tier companies make no such attempt.

8.5.1 Recruitment

Japanese companies recruit at least 90% of their engineers directly from universities. Consequently, great emphasis is placed on keeping good relations with professors; also, graduates employed by the company are very active in recruiting new engineers at their Alma Mater. Some universities try to restrict the number of their graduates that go to a specific company. Also, many universities have a professor in each department who has control over all recruiting for that department.

In Japan, there is a specific day on which all companies can begin to recruit new engineers. In recent years, Japanese companies have regularly been jumping the gun by making offers to the best students before the agreed-upon day.

Japanese companies do not emphasize grades when recruiting. While these are of interest, personality – particularly the ability to work together with others – is more important. In looking at the entire Japanese system, one can make an educated guess as to why this is so. Every graduate of an engineering school in Japan has successfully completed many years of competitive schooling and passed a highly competitive exam *before* entering the university. Consequently, Japanese companies can be secure in the knowledge that every Japanese engineering school graduate they recruit has already demonstrated the necessary qualities to succeed. Therefore, they have the luxury of concentrating their recruiting attention on other attributes of future employees.

The Japanese cannot recruit sufficient numbers of certain engineering disciplines. Many companies have training programmes to give, for example, mechanical engineers those additional skills required to do electrical engineering work. It is interesting that people who have just spent four or five years studying a particular field are willing to shift fields shortly after starting their professional careers.

The remaining 10% of engineering recruits come from other companies, but first tier companies restrict the age of such recruits. Some traditional Japanese companies are willing to recruit a few specialists in their thirties, but most recruit only people under the age of 30. The average employee age of the company is one of the first statistics given by many companies. Because salary is tied directly to years of service, keeping the average age of the company low translates into relatively low labour costs. Also, the energy and vitality of a company depends in large measure on the average age of its employees. Second and third tier companies appear to have fewer restrictions on the age of their employees and many of them employ workers retired from first tier companies.

In 1986, Japan passed a law requiring equal employment opportunities for women. Nonetheless, there are still very few women engineers or managers. One reason could be that the law is new, so there are few women in engineering in Japan. Another possibility is the following. Some Japanese companies adopted a two-track arrangement for their professional staff after the law was passed. Track A is for those professionals who are willing to move while Track B is for those who are not. Most women choose Track B because they would like to have a family and feel that moving would be detrimental to that end. However, only those in Track A are eligible to enter management. When a new recruit enters a company with the track system, they must *immediately and irrevocably* choose a track. Consequently, such companies effectively exclude women from management while still meeting the letter of the law.

It is quite common for Japanese workers – primarily men – to live apart from their families for many years, with visits home at a frequency varying between weekly to several years (for workers overseas). This undoubtedly has some negative impact on children of workers, but it has one very good side effect from the cultural viewpoint. Their children are not moved from school to school and, more importantly, a strong connection is maintained with the child's extended family (grandparents, uncles, aunts, cousins, etc.). This assists the transfer of culture to succeeding generations.

The personnel department has the final say in all hiring and promotion decisions. While technical managers can request certain skills, it is up to the personnel department to decide whether the position will be filled and who will fill it. In manufacturing companies, the personnel manager is often an engineer. Consequently, hiring decisions are made by someone with a technical background. Also, since the personnel manager is rotated to another position after a few years, it is likely that he will be prudent in his decisions.

8.5.2 Training and Development

All companies have training and development programmes. These vary from on the job training to entire centres dedicated to training and development of employees at all levels. All companies emphasize the importance of continuous training and development and they invest heavily in such activities. As will be discussed below, once a Japanese engineer is employed by a manufacturing company, there is little chance that he/she will leave. Consequently, companies know with certainty that their investment in training and development will stay with the company.

At many companies, every new engineer is required to start on the shop-floor, working as a machinist or in a similar capacity. In addition, there is a specific effort made to teach all employees what the company does. One story often heard in Japan is that the Japan Railway Company requires all its new employees to take tickets at Shinjuku Station during rush hour so that they will know who the customer is and what it is like for the employees who are in contact with the customers. (Shinjuku Station is the busiest in the world with more than two million people passing through it every day.)

Both technical and business training is given to employees. For example, the Sanyo Electric Training Facility in Kobe has a comprehensive in-house training programme, and Boston University's School of Business teaches all courses (on-site) required to obtain an MBA. Japanese engineers working in manufacturing companies are at least as well trained in all aspects of business as US MBA graduates.

8.5.3   The Workplace

The workplace in Japanese manufacturing facilities appears very similar to that of US facilities. Such is not the case in Japanese research laboratories. For example, in the research laboratories of Minolta near Osaka and Sanyo Chemical in Kyoto, workers' desks are in the laboratory in an open arrangement. The Japanese feel that this promotes team work – a much sought-after characteristic of Japanese companies. In the Minolta Laboratory, the entire building has been designed to force researchers to interact many times each day. Rest rooms and the cafeteria are centrally located so that one must walk down a long hall past many laboratories (most with open or no door) to reach these facilities.

Further, large, expensive equipment, located centrally, is available to all researchers, *and* they are permitted to operate the equipment themselves. The companies believe that this promotes personal involvement in all aspects of research and is a "perq" for research staff.

Examples of other interesting aspects of the workplace include the following. A Fujisawa pharmaceutical manufacturing plant has a Shinto Shrine to which a priest comes monthly and all employees gather to pray for the safe operation of the plant. In many Japanese manufacturing plants, employees exercise together at set times during the workday. A Minolta manufacturing facility has a Japanese tea room and a Japanese garden used by employees to learn and perform the Japanese tea ceremony. Such examples demonstrate how Japanese industry incorporates various aspects of Japanese culture directly into the workplace.

8.5.4   Promotion and Personnel Appraisal

Personnel appraisal is a pro forma process because it is a foregone conclusion that everyone will be promoted at roughly the same age and that everyone will receive approximately the same raise each year. However, even though everyone of the same age (actually, the same number of years of service to the company) gets a new title at approximately the same time, some promotions include increased responsibilities while others include no new responsibilities. Those workers who receive additional responsibilities are on track to reach management. There is a pyramid structure to each company and only those people who reach management are eventually retained. Especially after age 40, increasing numbers of workers are transferred out of first tier companies into subsidiaries and to suppliers. This leaves the first tier companies with the best managers and a lower average age.

Only a very small number of employees have problems with alcohol and drugs. These employees are the responsibility of the immediate supervisor and the employee's peers. Much like a Japanese family, individual work units compose an "inside" group. All members of this group are helped by each other. (Even deaths in the family, births, weddings, and other personal events provoke assistance by this work unit.) Consequently, the central personnel office rarely encounters problem employees.

It is interesting to note that there are no "Employee Assistance" programmes in Japan. Indeed, the largest bookstore in Tokyo does not even have a *section* on Psychology.

8.5.5   Compensation

All Japanese college graduates start out at approximately the same salary, irrespective of field. However, over the years, salaries in certain fields increase more rapidly than in others. Manufacturing companies give engineering graduates with a Master's degree credit for two extra years of experience, so an engineer with a Master's starts at a higher salary than had he/she started in the company immediately after completing the B.S. degree. (Recall that all engineering schools we visited said that students could obtain a Master's degree in one year. Consequently, an engineer with a Master's is a full year ahead of others in salary throughout his/her career. This leads to a large number of engineers in manufacturing who have a first graduate degree *in Engineering*.)

Salary is paid in two parts: base salary and bonus. In 1990, base salary for new engineers in Japan was approximately $20 000. In addition, a bonus of about $10 000 is paid. Additionally, most engineers work an average of two hours per day of overtime. Japanese engineers are *paid* for *all* of their overtime hours. This adds 25% to the base earnings of Japanese engineers or another $5000 per year.

First tier companies and government agencies (but not universities) provide housing subsidies. These include renting company-owned housing to engineers for as little as 10% of the going market rate and co-signing for housing loans so that extremely favourable interest rates are obtained. Additionally, the companies and government agencies pay for *all* commuting costs. Health and life insurance benefits appear to be similar to those found in the USA. (However, health costs appear to be under far better control in Japan than in the USA.) Retirement benefits seem to be adequate to maintain a basic lifestyle until government benefits begin around the age of 65.

Presidents of Japanese companies earn about seven times the amount earned by a new engineer. While there is some variation from company to company in this figure, the variation is not large.

8.5.6   Corporate Relationships

A very important aspect of compensation is the bonus. This varies from year to year depending on the profitability of the company. It is a crucial mechanism in maintaining the "lifetime employment" system. A Japanese company can reduce its labour costs by as much as 30% simply by not paying a bonus. In a downturn, this enables companies to keep their staffs employed.

If this is not enough, first tier companies reduce the profit margin of their second tier suppliers and second tier companies reduce purchases from third tier companies. Since employees of third tier companies are often housewives, working part time in their homes, when these employees have no work they do not show up on unemployment statistics. Other observations about the relationship between first and second tier companies follow.

At first tier companies one can commonly see trucks lined up in the street outside manufacturing plants with their engines idling. These are owned by second tier companies and hold parts needed by the first tier company. First tier companies practise "Just-in-Time" manufacturing, so they do not accept the parts until they are needed. This permits the first tier companies to keep their inventory costs down and improves their profits. However, the second tier companies always keep stockpiles of parts because they cannot afford not to deliver an order to their customer. It is not clear whether Just-in-Time manufacturing actually saves any money in the overall Japanese economy since the costs of storing spare parts is borne by someone.

Second tier suppliers maintain close social relationships with both their first and third tier partners. Friendly discussions several times a year, with alcoholic beverages available, play an integral role in these relationships. The importance of drinking alcoholic beverages with suppliers was pointed out by the Japanese plant manager of a large IBM manufacturing plant which we visited in 1990. He pointed out that his suppliers cannot understand why they cannot drink beer with him when they sit down for their discussions. IBM has a policy that no alcoholic beverages are allowed in any of their plants anywhere. The plant manager has tried unsuccessfully to get IBM to rescind this rule for Japanese plants.

The whole matter of first, second, and third tier companies is a reflection of the underlying culture. Taken together, they compose an "inside" group which is hard to break into. Many US companies complain that the Japanese will not let them have any business. It appears to us, however, that US companies have not been willing to become a part of an "inside" group. This would require moving nearby Japanese companies and developing relationships over many years.

8.5.7   Labour Relations

Japanese workers are highly unionized; however, there are few "Trade" unions. Rather, most workers belong to "Enterprise" unions. An Enterprise union is usually a group of people located at a single plant who band together to bargain with the plant management on wages and working conditions. Everyone at the plant – from secretaries to engineers – belongs to the union. This situation prevails until a person joins the rank of management. Many Senior Directors of companies have been officers of their union. Further, nearly every company we visited has at least one director who had been an officer in their union. This situation leads to a great deal of understanding between management and workers but does not result in complete acquiescence by either side. It does tend to temper excess by either side, and promotes consensus decisions.

Each spring, all of Japan's labour unions attempt to get management to increase wages. This event is known as the spring wage offensive. The unions tend to seek an increase based on the industry that is doing the most poorly. For example, if the steel industry has a bad year and can only afford to give a 3% increase, then all unions try to get 3%. Once again the concept of an "inside" group appears as all unions act so that their members can save face.

8.5.8 Small Group Activities

Quality circles are a well-known feature of Japanese manufacturing companies. These are small groups of employees who work together to improve the quality of a product by undertaking a series of projects based on their own experience. The Japanese use the same concepts to improve management, maintenance, and other things. The circles compete each year for company and national awards for quality.

The (Deming) national quality award is based on clearly defined standards which are set for the entire country by the judging body – an amalgam of Japanese professional societies. Any company can win at any time so there are many awards in a given year and companies that win continue to pursue the award the following year. Companies sometimes change the thrust of the quality circles after winning an award. For example, Toyoda Machine Works shifted from product quality to maintenance quality after winning the award. While Japanese companies are proud of winning the national quality award, they do not appear to use the awards in marketing. Rather, they use the awards as incentive to strive for another award.

Two issues are of particular interest: the similarity between these circles and the Japanese family and the fact that every company can win a national quality award every year. In some companies these circles take on the responsibility of making sure that neither its members nor the circle lose face. For example, many companies ask employees for suggestions to improve the work environment. These are given in prodigious numbers (50 000 per year is not uncommon). It is important that everyone make suggestions. If a member of a circle does not have a suggestion to give, then another member of the circle gives him one to put forth. As mentioned earlier, personal problems are also addressed by members of the circle. This seems quite similar to the way the Japanese family functions.

8.5.9 Non-regular Employees

The last feature of Japanese manufacturing we will discuss is the use of part time and foreign workers. Many manufacturing plants located in rural Japan employ part time workers. Toyota Motor Company recently announced the construction of several new plants in such areas. Local farmers are employed at the plants during the winter and work their farms during other times of the year. It seems likely that the culture and values of farmers are conducive to successful manufacturing [23].

Foreign workers are employed by a number of the companies, particularly in the automotive field. Automotive assembly work is among the most stressful types of manufacturing job. The relative lack of young Japanese workers willing to do such jobs probably explains this.

One last case that fits the mould of "irregular" worker is the American baseball-playing engineer employed at a large manufacturing plant. The young engineer was hired directly out of a US university because he is an excellent baseball pitcher. The company has a baseball team that competes in a Japanese industrial league. The company has a corporate philosophy that successful advertising is based on the number of lines of coverage of a company that appears in the newspaper. A successful baseball team gets many lines of print; hence the star pitcher.

8.6   Manufacturers in the USA have been Very Successful at Using Subcultures

There are a seemingly endless number of books which describe the American system of business. A typical example is Ref. 24. Briefly, American business can be grouped into three sizes: large, medium and small, as measured either in terms of numbers of employees or in annual revenues generated. These businesses currently tend to operate independently of one another. For many years, large businesses in the automative manufacturing field had preferred suppliers from the ranks of small and medium size companies, but in the past ten years or so this system has been changing rapidly. Another example is in the design and construction of power plants. Fifteen years ago it was common to have several billion-dollar plants ordered by a verbal agreement between a utility and an architect–engineering–construction company ón a sole source basis. Today, all such deals are in the form of complex, written contracts and must be put out for bid.

Small companies account for more than 50% of employment in the USA, with large and medium companies accounting for the remainder. When a person loses a job at any US company, they show up in the unemployment statistics.

8.6.1   Recruitment

American manufacturing companies recruit a relatively small percentage of their engineers directly from universities. However, when they do recruit from universities, the student's grades are of primary concern, particularly to the large companies. The probable reason for this can be seen as follows.

At the University of Pittsburgh, we have been educating engineers for almost 150 years. We have data that show that SAT scores are not a good predictor of success in our engineering school. However, doing very well in high school relative to your peers is an excellent predictor. Engineering school is highly competitive; it requires long hours of hard work as well as intelligence to graduate. As one measure of how difficult it is, not a single one of many thousands of graduates of the Mechanical Engineering Department over the past 40 years has achieved a 4.0 (straight "A") grade point average, despite the fact that many have gone on to become presidents of major, successful, US manufacturing companies.

Many US engineering graduates get into engineering school on the basis of SAT scores. Preparation time for these exams is relatively short (high schools commonly offer one year courses which prepare students for the SAT). Consequently, grades in college, earned in fierce competition with peers, are probably the best indicator of future success in the competitive world of business. As a result, US companies must emphasize college grades as the only available predictor of future success. (Many US companies emphasize other characteristics as well as grades but *all* companies emphasize grades.)

The recruitment of women and minorities is important to US companies because the pool of white males that has dominated the American manufacturing industry for the past 100 years is rapidly diminishing. A common method for the USA to increase the pool of technically trained people available to manufacturing companies is to change its immigration laws. The most recent change, passed last year, permits thousands of such people to enter the USA. This also dilutes the impact that American universities have on the manufacturing industry.

US companies actively recruit employees from competitors. As an example, when the author was employed in California's "Silicon Valley" during the late 1970s and early 1980s, an engineer was thought to be quite unusual if (s)he did not change jobs every two years or so. Until the early 1980s, most large manufacturing companies in the midwest and east of the USA practised a policy of "lifetime" employment similar to that of first tier Japanese companies. However, the perceived need on the part of these companies to reduce their average salary led to massive layoffs of older – presumably more highly paid – workers. These workers were often forced to move their families long distances to seek new employment opportunities.

The mobility of American workers, while not entirely new, has been increasingly common since the early 1970s when the number of women entering the workforce began to increase. Many children became displaced from their extended family and, with both parents working, children were often left with no adult models who had the energy and time to devote to them. A rapid increase in the divorce rate accompanied this phenomenon which also led to children having to grow up "on their own".

During the 1980s American business began to move its manufacturing to more rural climes. This exodus from the former major manufacturing centres was ostensibly undertaken for financial reasons. It appears to the author, however, that without realizing it, businesses were actually seeking employees whose personal culture was more mature.

If one looks at American industry from the cultural perspective put forth in this paper, it becomes clear that US manufacturing companies have thrived by using immigrant workers who were usually from rural settings in their country of origin. For example, the steel industry in Pittsburgh was built using peasants who emigrated from Europe. After several generations, however, their offspring forgot their rural roots. As a result, companies moved plants south to more rural areas of the USA or to rural areas of other countries like Mexico or Korea. In a sense, these companies "used up" the underlying culture. It is interesting to note that Japanese companies opening manufacturing facilities in the USA tend to locate them in rural areas.

In the USA, top managers of manufacturing companies often have no engineering training but many have MBAs. These people do have very high-level marketing, finance, and other business-related skills, but so do their Japanese counterparts. Yet, manufacturing is a technology-driven industry and technology requires mathematics, science, and engineering skills. As we noted earlier, it appears that even Japanese high school graduates have a better education in mathematics and the sciences than US college graduates. If this is so, then those US MBAs who do not have undergraduate degrees in mathematics, science, or engineering are inadequately prepared to compete with the Japanese in manufacturing.

Personnel departments obtain candidates for positions identified by technical managers but the managers make the final decision on who will be hired.

8.6.2 Training and Development

In the United States, training and development has historically been the responsibility of the individual worker. Most large companies do have training programmes for workers. However, these are usually rather general in nature, and are often done on the employee's own time. The reason for this situation is commonly attributed to the mobility of American workers. Companies feel that they will lose their investment if employees move and, perhaps more importantly, that employees will take trade secrets with them if they receive too much information.

Many engineers pursue graduate degrees on a part time basis, some in engineering, but many pursue MBAs. Technical short term training is offered by both professional societies and universities.

8.6.3 The Workplace

As mentioned earlier, the workplace in Japanese manufacturing plants is similar to that of American plants. However, there are marked differences in research and development facilities. Until the early 1980s, it was common to find outstanding centralized research labs in large US companies. These organizations produced many of the basic research breakthroughs that led to American manufacturing successes. During the past ten years, corporate R&D centres have been turned into profit centres. As a result, workers trained to perform research and development have been turned into entrepreneurs. The result has been a dramatic reduction in the number of patents issued to US researchers. This situation can be traced directly to the American educational system which produces large numbers of MBAs and the American industrial system that permits people trained in this discipline to control its management.

8.6.4 Promotion and Personnel Appraisal

American companies use many methods to evaluate workers. Except in unionized plants, promotion is normally done on the basis of accomplishment as defined by company management. As a result, one can find individuals of many different ages with identical titles. It is quite possible for a very young worker to have large responsibilities.

Employee assistance programmes have become quite common during the past 10 years. These programmes, paid for by the company, provide employees with access to professional psychologists and counsellors to help deal with personal problems. These programmes are a direct response by companies to help employees deal with the four existential fears identified by Yalom. Unfortunately, the programmes are restricted to a relatively small number of visits, so most people can only get "first aid" instead of a real cure. Nonetheless, these programmes clearly show that American industry is trying to grapple with the problem.

8.6.5   Compensation

Graduating engineers with a first degree currently receive approximately $30 000 in salary. The amount varies considerably depending on the particular field. Students entering non-engineering jobs usually receive considerably lower compensation. Pay rises vary from company to company and generally depend most strongly on individual performance. Most companies provide assistance in purchasing health, life, and disability insurance and many contribute to retirement programmes. Engineers are not compensated for overtime unless they are in a union shop. There is no general company assistance with either housing or commuting costs. It is interesting to note that the average salary for engineers does not vary greatly from region to region. For example, a few years ago, the difference in average annual salary for mechanical engineers between the highest and lowest cost of living areas of the US was only about $10 000. Since the average cost of housing alone differs by several hundred thousand dollars between these two regions, interest payments on housing would more than take up the salary difference.

There are usually three different rates of salary increase for technically trained people depending on whether they have a bachelors, masters, or doctoral degree. There are also often two different salary tracks along which a worker can move; a technical track and a management track. Since the management track is perceived by many technical people as the one which leads most rapidly to a higher salary, it is common for engineers to seek an MBA instead of a graduate degree in engineering.

The presidents of large US manufacturing companies earn about 40 times the amount that a new engineer earns. (In the past year, the highest compensation received by the president of a US manufacturing company was more than 2500 times that of a new engineer.)

8.6.6   Corporate Relationships

Rugged individualism extends to corporations in the USA. Antitrust laws are probably the historic reason for this, but the American idea that competition is more important than co-operation keeps it alive. Aside from professional societies, meetings between personnel from different companies are rare. At a recent conference on manufacturing [25], company representatives agreed that the most common route for information to travel from company to company is via salesmen who call upon companies and provide information of what is happening at other companies the salesman has called on.

Because of the competitive bidding methods that have become so widespread in recent years, it is difficult for companies to work with subcontractors who may also be bidding with a second, third, or more companies. Even when the government bends the antitrust rules to permit research on "pre-competitive items" by companies from a single industry, there is a wall between companies.

8.6.7 Labour Relations

Trade unions are relatively common in manufacturing companies in the USA. However, the number of employees who are unionized has been dropping steadily for some time. Trade unions are normally organized on a national level and negotiations usually conducted on a company by company – rather than plant by plant – basis. Membership in a trade union is restricted to individual workers who perform work of a particular nature. It is possible to have several trade unions represented in a single company.

Very few management people have been members of trade unions. Partially as a result of this, the nature of negotiations is usually quite adversarial. During the 1980s, the political power of the trade unions has diminished rather rapidly as whole unions have been destroyed by replacing striking members with non-union workers.

For non-union workers, including many engineers and managers, job retention has become problematic. Massive layoffs over many years have reduced – if not completely destroyed – feelings of loyalty to a company. The use of part time or contract employees to avoid paying for health insurance and other fringe benefits has become common. As a result, relations between companies and non-union workers are also quite strained.

8.6.8 Small Group Activities

During the 1980s, a number of US manufacturing companies attempted to copy the Japanese by establishing "quality circles" in their plants. The results have been mixed at best.

The US national (Malcolm Baldrige) quality award is given by the government (not a peer group), and winners use the award as a marketing tool. Only a few companies can win the award each year.

There have been several well publicized attempts to copy Japanese quality circles in the USA. In light of the apparent close connection between the underlying Japanese culture and their success in quality circles, one wonders whether such circles can work over a long time in the USA. The underlying culture in America tends toward rugged individualism, many American families are "nuclear" in that they are a long distance from grandparents, uncles, aunts, etc., and a great number of our children are raised in single-parent families. It seems that the challenge the USA faces in manufacturing is how to use its own culture successfully in structuring its manufacturing enterprises rather than copying methods that succeed in foreign companies.

8.6.9   Non-Regular Employees

As mentioned above, the use of part time and contract employees has increased rapidly during the past ten years. Even managers and executives are sometimes hired on a short term contract basis. The primary reason for this is the desire of top management to reduce overall costs with no consideration for the long term impact on the company's ability to remain technically competitive in manufacturing.

8.7   To be Successful, a Manufacturer Needs to Fill In the Gaps in Its Employee's Culture

It should be clear from the preceding discussion that culture plays a role in determining success in the field of manufacturing. But even if one accepts this thesis, the question remains: "What can be done to take advantage of strengths and correct weaknesses in the underlying culture of workers?". How does one "deal with" an existential anxiety? In one sense, there is no answer to this question. Yet we all experience times in our lives when we feel quite comfortable, so there must be some good techniques available. This paper opens up the question, but considerable work is needed to find comprehensive solutions for companies. Nonetheless, a few things seem likely to be of help.

Selecting workers who already possess a mature underlying culture is the most obvious solution. Many companies are doing this without even realizing it. For example, I brought these ideas to the attention of an executive of a major American communications equipment manufacturer. His company had recently selected a manufacturing plant site in a rural region of a European country. However, the company had not recognized the importance of obtaining workers with a mature underlying culture. The more they thought about it the more the company began to see that this had indeed been a strong reason for the choice of location.

The majority of workers in most industrial countries now or will soon live in large cities. Consequently, simply locating a plant near a supply of workers with a mature underlying culture is not always possible. In situations like this, one approach would be to restrict management positions to individuals with strong technical backgrounds who have been given sufficient management training, and who are committed to dealing with existential anxieties in their personal lives. This would require additional training and expenditures on the part of the company but the pay back could be rapid and large.

Extending employee assistance programmes to permit long term work is another possibility. If individual companies found this economically unfeasible, they could push for federal legislation to support such programmes.

One last suggestion, companies can introduce new types of training that begin to help employees deal with existential anxieties. On a practical level, one of the best "handbooks" ever written on this topic was published in 1959 [26]. The book, entitled *The Management of Time*, was written by James T. McCay. It is not a book for the faint hearted. The ideas McCay puts forth require tearing down of defences built up to help deal with anxiety so that one can learn how to deal more directly with it. For example, he attacks habits as destroyers of freedom and suggests several simple but creative ways to help see how difficult it is to change habits. He suggests that experiences are born in the brain and that breaking down those experiences to finer and finer detail is a road to alertness.

There are, of course, numerous books written on the use of psychology in the management of people. These books tend to be encyclopaedic, which leaves the manager with an overdose of information, but no clear understanding of the problems. This chapter suggests a relatively simple, yet firmly grounded, road map for managers in manufacturing companies to follow. Whatever path a company chooses, it seems clear that ignoring the underlying culture of its employees holds perils which none dares risk.

References

1. Gunn TG. 21st century manufacturing. Harper Business, New York, 1992
2. Stephanou SE, Spiegl F. The manufacturing challenge. Van Nostrand Reinhold, New York, 1992
3. Kotter JP, Heskett JL. Corporate culture and performance. The Free Press, New York, 1992
4. Miller LM. American spirit: visions of a new corporate culture. W. Morrow, New York, 1984
5. Weiss JW. Regional cultures, managerial behavior, and entrepreneurship: an international perspective. Quorum Books, New York, 1988
6. Carnevale AP. America and the new economy. Jossey-Bass Publishers, San Francisco, 1991
7. Drucker PF. Managing for the future. Truman Talley Books/Dutton, New York, 1992
8. Webster's ninth new collegiate dictionary. Merriam-Webster, Springfield, MA, 1986
9. Based on the author's personal notes and interviews gathered while engaged at the Three Mile Island nuclear power plant during April and May, 1979
10. Schein EH. Organizational culture and leadership. Jossey-Bass Publishers, San Francisco, 1985
11. Becker E. The denial of death. The Free Press, New York, 1973
12. Becker E. Escape from evil. The Free Press, New York, 1975
13. Roheim G. The psychoanalytic interpretation of culture. In: Muensterberger W (ed) Man and his culture: psychoanalytic anthropology after 'Totem and Taboo'. Rapp & Whiting, London, 1969, pp. 31–51
14. Graves D. Corporate culture – diagnosis and change, St. Martin's Press, New York, 1986.
15. Yalom ID. Existential psychotherapy. Basic Books, New York, 1980
16. Kolar MJ. Culture of manufacturing engineering in Japan: towards improving the international dimensions of engineering education. Project report submitted to the US Department of Education, Fulbright – Hays Group Study Abroad Program, 1991 (also available from the Department of Mechanical Engineering, University of Pittsburgh)
17. Hendry J. Understanding Japanese society. Croom Helm, London, 1987
18. Smith RJ. Japanese society – tradition, self, and social order. Cambridge University Press, Cambridge, 1983
19. Spindler G, Spindler L. The American cultural dialogue and its transmission: The Falmer Press, London, 1990
20. Okimoto DI, Rohlen TP (eds). Inside the Japanese system – readings on contemporary society and political economy. Stanford University Press, Stanford, CA, 1988
21. Fields G. From bonsai to Levi's – when west meets east: an insider's surprising account of how the Japanese live. New American Library, New York, 1983.
22. Inohara H. Human resource development in Japanese companies. Asian Productivity Organization, Tokyo, 1990
23. This idea was suggested to me by Dr. Dwight Baumann of Carnegie Mellon University in 1990

24. Musselman VA, Jackson JH. Introduction to modern business, 9th edn. Prentice-Hall,
    Englewood Cliffs, NJ, 1984
25. Proceedings of the 1st annual Pittsburgh manufacturing systems engineering conference,
    Manufacturing Systems Engineering Program, University of Pittsburgh, 1990
26. McCay JT. The management of time. Prentice-Hall, Englewood Cliffs, NJ, 1959

9 Culture of Manufacturing – a Case Study: Approaching an Understanding of Different Production Styles in Japan, the USA and Germany

E.F. Moritz

9.1 Technology as a Social Process? Overall Views of the Effects of Culture on Production

9.1.1 Empirical Investigations Rather than an Eclectic Digest – the Methodological Foundation of This Contribution

Announcing a case study on the "culture of manufacturing" is a challenging claim – neither is the term yet satisfactorily defined, nor is it a self-explanatory constant entity which would simply need some illustration. Denomination, exemplification and comparison will therefore have to be carried out simultaneously to convey some feeling for the complex interdisciplinary connotation of the "culture" of manufacturing on the theoretical level, and this discussion supported with some practical characterizations of important features as well as of some peculiarities of individual production styles.

The foundation of the subsequent discussion was laid in standardized interviews and complementary discussions and factory tours with senior design managers in eight machine tool companies in Japan, the USA and Germany, respectively. This empirical comparative approach, plus the fact that the interviewees were still engaged practically in design work on the one hand, and were interviewed in their native language on the other, ensures that the results of this investigation are not just a compilation of PR strategies or a reiteration of cultural prejudices.

Going off the beaten track involves a number of dangers, however. The analytical and interpretative skill of the researcher is very determinative in the illustration of this complex subject, which may sometimes lead to an incomplete or distorted picture. Thus even though the empirical investigation was supplemented with a few intensive case studies, a large number of expert interviews, and a constant discussion and research co-operation with scholars from various disciplines working on a related topic, the reader may still have some additional comments to append or some criticism to voice.

In this case the author would appreciate direct feedback to improve his further research; he is naturally willing to share his opinions with other scholars as well.

9.1.2   Separation and Categorization – an Intellectual Historical Burden Impeding Effective Production, at least in the Western World?

"Culture of Manufacturing", "Fraktale Fabrik" [1], "Systemic Rationalization" [2] – all these recently coined catch-phrases point in the same direction, bluntly designated by the "I" in "CIM" – Integration. This is not necessarily a breathtaking new discovery, nor does the idea of integration at first glance seem to need some further lobbying. Is there not a lot of research being carried out to develop versatile unifying interfaces, MAP, LAN, WOP, CAQ, as industrial computer scientists would fluently inform you? Is integration, at least since the "discovery" of lean production [3], not a self-evident goal for every concerned employer or employee at whatever hierarchical position? It is not! Engineers not having the faintest idea about marketing (the reverse seems to go without saying) [4], CIM implementations getting all kinds of tricky but needless procedures out of the mainframe while neglecting a few simple features that would support the work force [5], and investment strategies attempting to make up the next quarterly report more than being concerned about the future of the company [6] – all these still constitute a common state of the art. Still worse, more often than not this situation is maintained without even a glimpse of absurdity.

One unfortunately valid example from the case studies starkly illustrates this disrespect of integrative thinking. The assistant to the manager of the department of business affairs in a (nevertheless) quite successful German machine tool company, after having smartly completed the "standard" factory exposition tour, was supposed to take me to the design department. He hesitated, and directed himself to the head of the design department. "Where is this thing?"

To be sure, this phenomenon is in no way confined to business managers. Similar examples of exaggerated intellectual categorization may be found in all other fields of science and engineering, at least in the Western world. This observation is more important to the discussion of the culture of manufacturing than one might think at first glance – for two reasons. First, many of the current debates and critiques of existing management practices imply that the solution to obvious "structural problems" may simply be a reshuffling of personnel, the adaption of a few lean or mean modes of administration, and an improvement of computer interfaces. Necessary as these may be, they are in no way sufficient. Putting it more dramatically: lacking an appropriate philosophical foundation – an "integrative attitude" to back up the organizational changes – these changes will be subject to constant re-evaluations and retractions, and they may finally absorb even more energy and finances than they release. Tearing down the intellectual barriers between the different departments or disciplines must therefore be one main objective of any discourse on the culture of manufacturing. The second reason to devote some consideration to the ideological encrustment in Western industry is still more practical and of urgent interest. Whereas the considerations outlined above had slowly merged their way into the theoretical discussion on a rather abstract or scientific level, Japanese companies actually demonstrated quite clearly and effectively that over-wrought abstraction, specialization, and intellectualization are in no way necessary preconditions to "worldly" success, at least not in a capitalistic environment.

How could this happen? Why can the Japanese take the advantage while simply ignoring some of the most adored attributes of Western intellectual evolution – systematic abstraction and innovative modelling? The answer to this phenomenon requires a few theoretical reflections. Philosophical, especially metaphysical insights and attitudes, quite naturally follow the perception of the world generated by physicists, "common sense" still trailing behind by a few generations. Western common sense, including engineering practice as well as most scientific modelling, for the most part is still influenced by the Newtonian physics which postulates that the world can basically be described and future events calculated by a comprehensible number of equations. Correspondingly, all phenomena in industry also appear to be eligible for elaborate modelling and thereby to be comprehensively manageable, no matter which aspect of administration or production is concerned. Therefore, a company is perceived as a clustered artifice of departments, vocations, and hierarchies, rather than as an integrated universe dedicated to materialize production.

Since the beginning of this century quantum physics has gradually disclosed a quite different reality. The further investigations attempted to penetrate the last (meta)physical miracles, the more complex the universe revealed itself to be. In fact, it turned out to be much too complex to continue to support the idea of putting the whole world into a few elegant equations. A lot of phenomena just do not fit into the established models; even simple particles do not like to be tracked down exactly at a point in time, or they may simply refuse to exist despite a logically "sound" theoretical approach.

The "new picture" of a world being on principle not completely intelligible has since gradually penetrated into other disciplines as well, unfortunately still not much further than into nonlinear dynamics and mathematical fiddling. There does not seem to be an understanding that a company is also a highly nonlinear chaotic construct and is quite unwilling to function according even to elaborate models. Most managers still insist that product marketing or manufacturing organization can be comprehended and administered exactly, if only the conglomeration of defining equations were elaborate enough.

The Japanese, who historically have not bothered too much about abstract cognizance anyway, intuitively put into practice what Western intellectuals are still struggling to comprehend. If the universe on the micro as well as on the macro level is a chaotic mess rather than a structured interrelation of equations, it makes little sense to base all management decisions or engineering thinking on elaborated theories either. Adaptable specialized structures on the organizational side, and models, methods and hypotheses on the intellectual, do have some practical justification and, employed perceptively as supportive tools, are certainly of great help in some instances. However, they just as often will prove to be wrong, not applicable, or at least not suitable. Pig-headed attempts to understand and stick to them as something more than tools have often ended up, as some examples in the succeeding sections will demonstrate, being quite disastrous.

The "Japanese alternative" is simply to promote awareness instead of intellectual arrogance. The production process is thus flexibly organized and iteratively improved to serve the intended goal instead of being strictly aligned to some foggily derived management theories. This sounds simultaneously plausible to the pragmatic thinker as well as intellectually inferior to the "Newtonian highbrow". However, Western business managers, in addition to improving their "own way", will seriously have to ponder about their intellectual foundation as well if they want to keep their engineers from becoming the creative court jesters of the Japanese company presidents before too long.

Specialization does not mean segregation. If managers and engineers finally come to appreciate this lesson and to foster their interdisciplinary awareness, they will greatly increase overall efficiency as well as job satisfaction, and will probably be on the right track towards a true "culture" of manufacturing.

Promoting awareness, with respect to the complexity of production and the manifold parameters determining the culture of manufacturing, is also the main objective of Fig. 9.1. The unfathomable interplay of mutually influencing factors should furthermore reinforce the idea that, if at all possible, quite a number of disciplines will have to co-operate to meaningfully advance upon a more complex understanding of production.

9.2 Contrasting Production Systems in Different Countries – a Playground for Misconceptions and Speculations?

International comparative research reveals some most striking examples of exaggerated intellectual categorization. The "success" of a nation in the capitalist world (regrettably the only or at least the most common standard by which countries are evaluated and ranked) is, depending on the individual point of view, attributed to trade practices, exchange rates, the size and formation of conglomerates, working hours, lot sizes, the degree of automation, the educational system, the structure of family, and whatever else somebody may come up with. Little does it seem to matter that most of those factors are somehow related to one another; i.e. that a high degree of automation still only makes sense in the production of larger lot sizes, that a hierarchical structure of the family and a certain set of individual values are prerequisites for longer working hours, and that trade practices and exchange rates not only represent the current economic necessities, but also mirror the aspired image of a country plus a good deal of strategical considerations.

One additional aspect has usually commanded even less consideration, although it should be a matter of course in analysing different cultures: the difficulties of comparing concepts and categories derived from contextually and structurally different systems. To name a few examples: patent statistics are usually referred to without mentioning the substantial deviations in the respective legislation; education is evaluated only according to the number

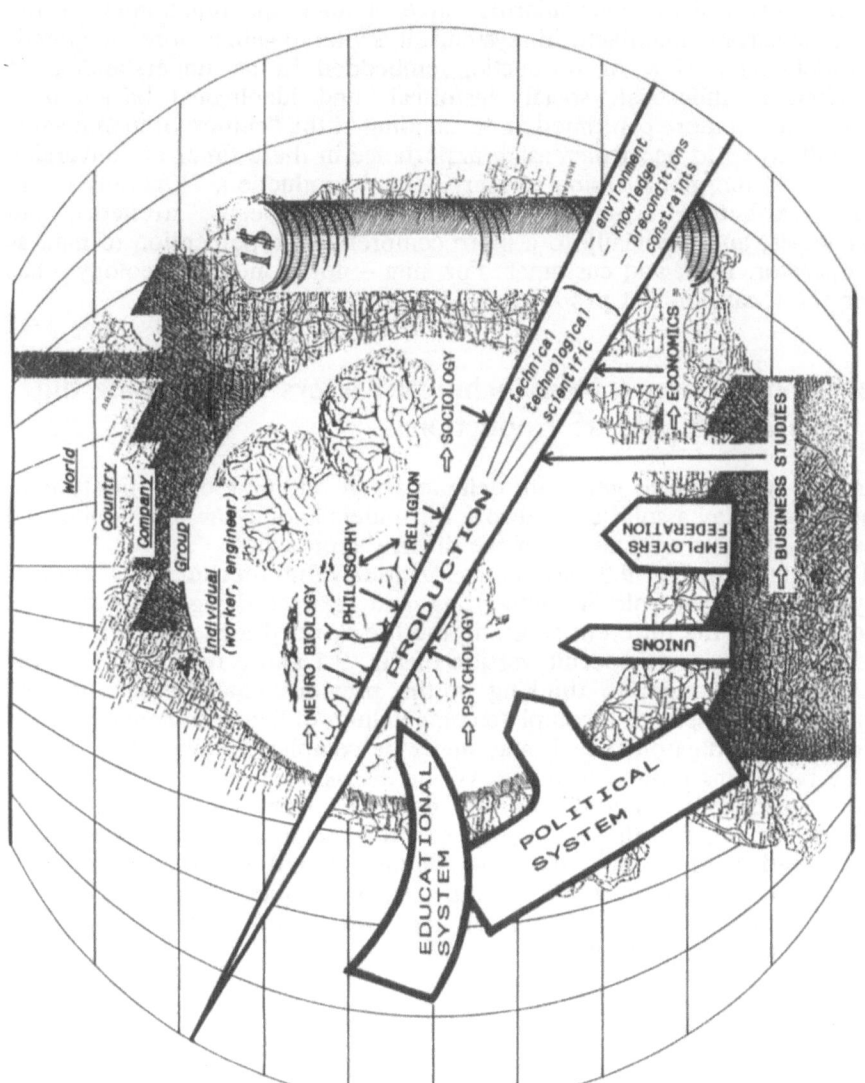

Fig. 9.1. The amalgamation of concurrent factors influencing production.

of school-years attended;[1] and the duration of innovation cycles in industry is compared without considering whether it is an overall new product development or a conglomeration of minor modifications.

These considerations once more illustrate and reinforce the starting point and the main objective of this contribution. Instead of desperately clinging to coercively isolated particularities in explaining the functioning or the success of certain manufacturing systems, it is vital to gain a more integrated, interdisciplinary view of production embedded in an understanding of the complex industrial, social, historical, and ideological background. Approaching a more conjoined understanding of the "culture of manufacturing" will prove to be of increasing importance in these times of converging nations and the globalization of markets and production. Hopefully it will lead to a better recognition of one another's needs, strengths, and weaknesses, and especially to a more comprehensive perception of man as the operator, user, and customer. For man – not money or ideology – has to be the focal point of production.

9.3 Step by Step – Non-Technical Factors Determining the Various Stages of Production

It is now high time to get more definite about outlining different styles of production to lay a solid groundwork for understanding the functioning and the strengths and weaknesses of the diverse approaches.

To begin with, Fig. 9.2 shows a compilation of non-technical factors that play an important role in influencing and characterizing the individual "moulding" of the different stages in the life cycle of a product. It may be interpreted as a more specific version of Fig. 9.1 and a tribute paid to the exaltation of integrative thinking in the previous chapters, and should certainly not be seen as a complete compendium of "environmental" factors influencing production (which may never be completed anyway).

The two items printed in italics will be discussed in greater detail in the following sections using results and examples from the case studies as well as some interesting outcomes of other related investigations. This representation of a few interesting factors should be understood as a demonstration of the postulated complexity of production; it should not be misinterpreted as a revelation of the two secret keys to the final comprehension of production styles.

---

[1] Ninety percent of the Japanese finishing the *kotogakko* (high school) and 75% of the American students completing the high school in some studies appear to constitute a huge intellectual advantage as opposed to just 25% of German pupils having achieved the *Abitur*. Factors like the educational system, the contents of study, the pedagogical style, the general philosophical attitude towards learning and knowledge (even in the "West" substantially divergent as a result of different interpretations of enlightenment), and the quality and educational background of the teaching personnel are generally nonchalantly neglected.

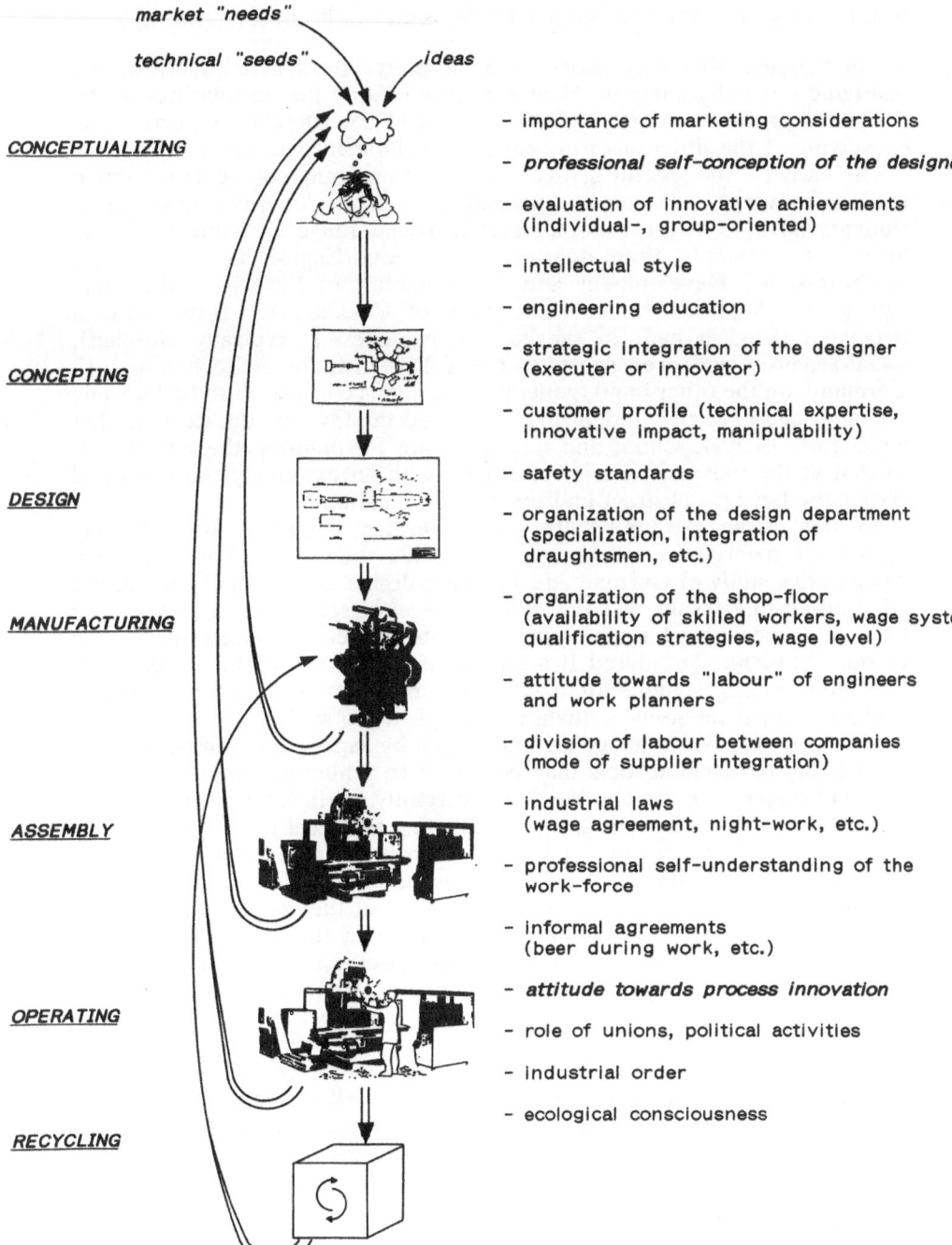

market "needs"

technical "seeds"   ideas

CONCEPTUALIZING

– importance of marketing considerations

– *professional self-conception of the designer*

– evaluation of innovative achievements
  (individual-, group-oriented)

– intellectual style

– engineering education

CONCEPTING

– strategic integration of the designer
  (executer or innovator)

– customer profile (technical expertise,
  innovative impact, manipulability)

DESIGN

– safety standards

– organization of the design department
  (specialization, integration of
  draughtsmen, etc.)

MANUFACTURING

– organization of the shop-floor
  (availability of skilled workers, wage system,
  qualification strategies, wage level)

– attitude towards "labour" of engineers
  and work planners

– division of labour between companies
  (mode of supplier integration)

ASSEMBLY

– industrial laws
  (wage agreement, night-work, etc.)

– professional self-understanding of the
  work-force

– informal agreements
  (beer during work, etc.)

OPERATING

– *attitude towards process innovation*

– role of unions, political activities

– industrial order

– ecological consciousness

RECYCLING

Fig. 9.2. Some examples of non-technical factors determining the various stages of production.

9.3.1   Case 1 – the Designer's Professional Self-Understanding

At first glance this may seem to be a pretty unexpected feature, not justifying a lengthy analysis. However, appreciating the peculiarities of the designer's professional self-understanding is of considerable importance in conceiving of the differences in product development strategies.

The basis for the specific self-conception of maturing engineers is formed in their university education. Japanese as well as American engineering students have to cover a comparatively broad range of subjects. These include, for example, three quarters in Freshman English and one quarter in Professional Development and Fundamentals of Speech at the Ohio College of Applied Science (University of Cincinnati); or two foreign languages (English and, in the case of engineering, typically German), social sciences, and even sports at the Tokyo Institute of Technology. In Germany, on the other hand, general education is considered to be basically accomplished once the student has completed the Gymnasium (a somewhat elitist form of high school and a prerequisite for entering the university), so that at the university he is confronted with an extremely thorough and systematic but also abstract and one-sided education.

So right from the university level, German designers are relatively competent specialists, and they also perceive themselves that way. In a comparative study of German and Japanese design education [7] one of the most interesting results was the different level of self-confidence expressed by senior mechanical engineering students. Whereas almost all of the German students considered it a matter of course to be able to design a swivelling milling spindle for a company, not even half of the Japanese students judged themselves qualified enough to do so.[2]

This lack of self-confidence exhibited even by Japanese graduate students specializing in machine tools may be traced to a number of roots of quite different origins. On the one hand, it is certainly a reflection of the different objectives of university education – cultivating technical understanding in a general context for further pragmatic refinement through the extensive company-specific training (in Japan) versus creating a devotee to a certain profession (in Germany), outlined above. One of the consequences of this difference in attitude is naturally a divergence in the qualitative level of individual abilities. Skills to tackle a quite specific technical problem like this, for example, will be less developed in Japan than in Germany, which of course also limits the self-confidence of the Japanese students in this respect.

A second phenomenon unequivocally influences the designer's self-conception as a whole much more than just his self-confidence. There is a marked difference in the approaches to master technical problem solving even at the university level. Japanese engineering students are basically familiarized with the adaptation, calculation, and optimization of a given

---

[2] Regrettably this study has not yet been extended to include American students as well. However, a subsequent survey encompassing all three countries has already been conducted and will be published in the near future.

concept; methodological training, if at all, consists of expounding straightforward procedures on how to develop a certain product.

On the other hand, German design education focuses on how actually to develop a concept that fulfils certain functional requirements utilizing a given set of basic machine elements. Methodological schooling here means instruction in systematic abstraction and analysis of given problems, as well as the heuristic creation of potentially appropriate solutions. In this context it is interesting to note that American universities favour still another approach. Although design methodology is undergoing some kind of revival after having been neglected for almost 20 years,[3] the integral part of design education consists of carrying through a whole design project from the very first stages of planning and conceptualizing to the testing and presentation of the final product. This practical, integrated, and comprehensive view of the design process appears to be the ideal preparation for industrial design practice; however, co-ordination and co-operation between universities and companies, at least in this respect, is quite meagre. Rather than being integrated into concrete product development projects, the students more often than not end up doing something like forging a toy fusion plant utilizing the resources of a remote village in Northern Siberia, which can no longer be considered an ideal training for the needs of design departments in industry.

A third factor influencing the degree of individual self-confidence as a whole is the character of the social environment – there is generally much less exhibition of self-confidence in typical "Eastern" cultures than in "Western" societies. A number of arguments attempt to explain this phenomenon. The main line of reasoning would be a social "conditioning" by a rather positive or rather negative feedback towards expressing self-confidence; which again is largely determined by the evolution of interhuman behaviour patterns in each country. According to some recent psychological studies [8], however, a different degree of self-confidence may also be interpreted with a different "construal" of the self as more interdependent in Japan and more independent in the USA and Europe.

Various empirical studies support this observation. Markus and Kitayama [8], for example, presented a questionnaire to both American and Japanese university students asking them to evaluate their own intellectual abilities in various fields as compared to the whole student body. American students in this case displayed a significant level of "false-uniqueness" considering only an average of 30% of their fellows better than themselves in fields like memory, athletic ability, and sympathy, whereas Japanese students on average hit the "correct" 50% mark quite correctly.

By now the professional self-understanding of the German designer has already been characterized quite completely. He has been thoroughly trained to master "engineering thinking" and technical problem solving and will

---

[3] After having gained the technological control over the Apollo programme (and thus recovering from the Sputnik shock) in the late 1960s along with the advancement of computer technology, design research in the United States appears to have been reoriented to emphasizing on application of hard- and software rather than brainware. This development influenced the design education into the same unfavourable direction as well.

then enter the industrial world in some kind of engineering profession typically feeling like an irreplaceable specialist no more than a few months after clocking in the first time. Once he has chosen to be a designer, it is hardly likely that he will ever change this vocation again – he would rather change the company, the type of business, or even the mode of employment – unless he is offered a chance to advance his career by becoming design manager or even managing director of the whole company.

In the USA, the designer's self-conception is less determined by professional training alone, even though he also certainly takes some pride in his profession. Unlike in Germany, however, the domination of capitalism as the utmost form of economic system is much less disputed. As a result, designers as well have much less unreflected consciousness of themselves being at the mercy of economic circumstances, which means that they subordinate themselves to the directives from the business management much more readily. Nor do they insist nearly as often as their German counterparts on their expertise as the incontrovertible foundation of any further proceedings.

For Japanese designers university education plays a much less significant role; their self-conception will at most be influenced by the rank of the university attended, but certainly not by the field of specialization. One example should illustrate this point. Only a very small percentage of the undergraduate or Master students from the quite renowned machine tool laboratory of the Tokyo Institute of Technology actually start working in a machine tool company. Typically students from a prestigious Japanese university choose to work for a similarly prestigious Japanese *kaisha* (enterprise) like a car or an electronics company, or, not uncommonly even for graduates from mechanical engineering departments, a bank. The whole machine tool business, due to the still comparatively small size of the companies and the lack of popularity of the product itself, is contrary to divergent beliefs in Western countries not considered prestigious in Japan. (However, it is nevertheless considered important!).

Yet the utmost, all-embracing determinant of the self-conception of the Japanese designer is the company, his *kaisha* [9]; as a matter of fact, this holds true for all Japanese employees of bigger, functionally self-sustaining companies.[4] This is also reflected in the designer's perception of the influence of the company's policy on his individual working style (Fig. 9.3).

Naturally the German "specialist" will not allow or at least not admit too much external influence on his working habits, whereas the Japanese "companist" conceives of himself more as an integral part of his company and thus as a functional element dedicated to performing according to the particular current necessities of his *kaisha*.

---

[4] Contrary to the widespread claims maintained nowadays, this is neither "natural" for the Japanese people nor a direct consequence of Confucian philosophy. It was rather an extremely smart uniform policy by Japanese companies and especially conglomerates to transform the Confucian values of learning and loyalty to the family to working and loyalty to the company in the early 20th century.

How much is your professional work influenced by your company's policy?

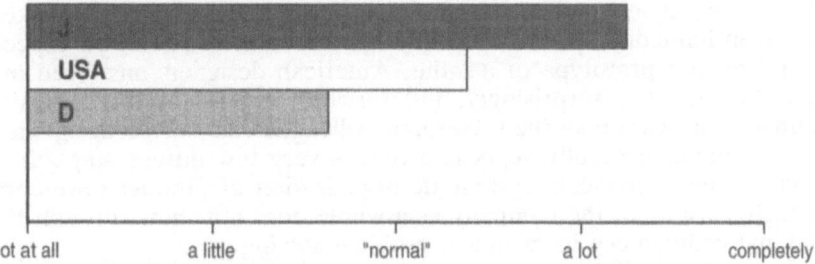

Fig. 9.3. The designer's perception of the influence of his company's policy on his own working style (mean value as expressed in the interviews).

9.3.2 Consequences of the Designer's Self-Conception on His Working Style

The professional self-understanding of the designer – as a self-confident specialist in Germany, a pragmatic tackler in the USA, and a flexible executor in Japan – exerts a substantial influence on the working style of the engineer. A Japanese designer is ready to be at the almost unlimited disposal of his company; he may formally be integrated into any meaningful working group, and he will try to be a co-operative member to serve the particular needs of this group as well as the overall goals of his company. This idyllic state of affairs is not without drawbacks, however. Being broadly educated and flexibly utilized, a Japanese designer has little chance to build up domain-specific expertise in one particular field. Along with his organizational mode of integration and his social background of reacting instead of acting, this also results in comparatively little instant competence in technical problem solving.

Interviews with Japanese designers revealed an interesting contrast between, on the one hand, a comprehensive integrated knowledge of the company-specific proceedings, necessities and potential problems in the overall organization of product development, versus, on the other hand, considerable difficulties in proposing *ad hoc* strategies to tackle a specific technical problem.

Product development process plans of one form or another are utilized quite extensively in all of the surveyed Japanese machine tool companies, although these are more pragmatic and less uniform than most Western "experts" in this field would suspect. As demonstrated in an earlier study [10], Japanese designers generally appear to have a more integrated view of their job, acknowledging the importance of investigating the customers' needs and taking into account manufacturability and simplicity in assembly much more seriously and effectively than in Germany (and, as revealed in the current study, in the USA).[5]

---

[5] The negligence of manufacturability by the designers may, at least in Germany, also be accounted for by the increased division of labour in product development. Unlike in Japan, German designers generally produce only a refined sketch, and leave the further elaboration to the draughting personnel. In the United States, however, the increased application of CAD at least in some companies resulted in a dismissal of engineering draughtsmen and an assumption of the production of detail drawings by the designers themselves. Nevertheless, these still did not seem to bother too much about the succeeding stages of production.

Confronted with a concrete technical problem, however, the responses of the design engineers disclosed a diametrically opposed picture. Asked to propose an immediate procedure to design a feed drive for a new especially fast and precise prototype of a lathe, American designers answered in the greatest detail and, surprisingly, put forward at least as methodical and systematic an answer as their German colleagues. Japanese designers, on the other hand, generally suggested only a very few diffuse steps to fulfil this task – the convincing systematic organization of product development obviously existing in the company as a whole does not show through to the individual competence in technical problem solving.

This surprising dichotomy between a sound understanding (and, as the success of Japanese companies clearly demonstrates, effective transformation) of group-related proceedings and at the same time rather meagre individual skills in pragmatic concept design may certainly not be attributed to the designer's self-conception alone. However, the perception of the role of the self and its integration into the specific structure of the social environment are important and in most of the current management-biased studies neglected parameters in comprehending and illustrating an integrated view of different production styles.

A closer examination of the renowned design proficiency of German engineers reinforces this argument. The ability to abstract technical contexts and to apply scientific methods as well as profound expertise in design are undoubtedly important prerequisites for achieving leaps and bounds in domain-specific innovation; but here as well certain limitations to this glory must not be overlooked.

A comprehensive series of questions concerning different aspects of technical problem solving [7] revealed that German engineering students are superior to their Japanese fellows (both groups were in their final year at the university) as far as the synthesis and development of pure mechanical solutions was concerned. Japanese students, however, offered a wider range of possible solutions taking into account chemical and even biological solutions to a technical problem. Interestingly enough, the technical blinkers worn by the German students did not seem to be an intrinsic property; just if the type of question somehow looked like suggesting a mechanical solution, they appeared to be victims of their one-sided education and generally stuck to their "expertise" in mechanical design rather than being able to produce ideas from fields foreign to their subject.

Product development in industry also offers more than enough examples of the "mechanical bias" of German engineers. A description of the early days of NC machine-tool design proves that this phenomenon is all but a new one: "Certainly mechanics as well as electronics of the machine tools have improved considerably. The most important area, however, the unity of machine and control, still too clearly exposed the 'link'. They designed controls for the machines instead of machines for the controls" [11].

Self-importance instead of self-confidence, in connection with misunderstood specialization fostered by the intellectual history of over-categorization, lead to an exaggerated segregation in product development as well as in the internal structure and administration of companies (and universities). This aspect unfortunately neutralizes many of the cherished strong points in German industry.

9.3.3   Case 2 – the Attitude Towards Process Innovation

Process innovation, the adaptation and actualization of new forms of management and the application of new technologies in the various stages of production, is another interesting subject to demonstrate the amazing influence of non-technical factors on the "shaping" of industrial practice.

Again Japan shows a most fascinating and unexpected profile. Contrary to popular opinion, Japanese factories are not a real-life preview of the new "Tron" sequel. Whereas there is certainly an astonishing degree of overall automation, right in-between the flexible manufacturing system (FMS) on one side, and the high-precision computer numerically controlled (CNC) lathe on the other, there may well be a tottering milling machine of the post-war period. And, amazingly, it may be just the right tool for the machining purpose intended.

From the case studies, along with supplementary observations by other researchers,[6] the following picture can be drawn. Japanese companies, although willing and financially able to advance automation in manufacturing and assembly, have a strong strategical consciousness, a stunning organizational flexibility, and focus effectively on the most promising areas of process innovation. They will group together an appropriate "family" of workpieces irrespective of structural or organizational barriers to implement a most productive FMS, manufacturing the remainder either with high-precision CNC machining centres or with anachronistic engine lathes, whichever seems suitable. They will utilize predominantly German ultra-high-precision (and ultra-expensive) measuring machines and still do most of the quality control manually, and they will automate the supply of parts from the stock to the assembly station with unmanned carriers and then set up the machine tool by hand. The manufacturing area of Japanese companies may therefore rightly be described as *bricolage* [12] or *Bastelei* [13] (both conveying the meaning of a botched job); however, "fit for the use intended", the adored attitude of American production engineers, is certainly accomplished most effectively on the shop-floors of the Japanese machine tool companies.

Simple as it may seem, the Japanese can build their process innovation policy on a unique threefold groundwork of a dualistic industry, an intricate, performance- rather than job-dependent wage system, and an intellectual structure demanding a devotion to rules and regulations and allowing contradiction and inconsistency. All three factors have been widely discussed in literature, so here they will just be summed up briefly with regard to their respective influence on the Japanese automation strategy.

The separation of large and small companies in Japan is much more stringent than in the USA, and the tiny Japanese "sweat shops" can certainly not be compared to the skilled artisan supplier industry in Germany. They generally enjoy neither new technologies nor in-house engineering expertise,

---

[6] Enlightening expert interviews revealing different perspectives of process automation in Japan were carried out with Eugene Merchant (IAMS, Cincinnati), Klaus Dustmann (Gildemeister NEF, Bielefeld), Norbert Altmann (ISF München) and Masami Nomura (Okayama University).

but are patriarchally confined to bigger supplier companies and through these to the large firms dominating industrial life in Japan.

This segregation is stabilized recursively through some smartly evolved particularities in the Japanese industrial order (see also [14]). Bigger companies installed the system of life-time employment and the accompanying seniority principle to attract and keep skilled labour, and subcontracted unskilled work to small supplier shops. These were (and are) usually not powerful enough to introduce a similar system, so that they succeedingly lost skilled personnel and in the wake became increasingly dependent on contracts with the bigger companies, which again utilized their power to induce a firm pressure on manufacturing prices. As the small mills did not have the financial basis to substantially increase the degree of automation, they could only lower their costs by lowering the wages and exacerbating the harshness of working conditions, which in turn further diminished their attractiveness to potentially proficient employees. Even today the workers in small companies (10–99 employees) earn considerably less than their colleagues in large companies (more than 1000 employees); in the age group over 50 difference in salary amounts to more than 40% [15].

The pay policy in Japanese companies did not receive significant attention until recently [13, 16]; it nevertheless constitutes an important factor sanctioning the Japanese automation strategies. The salary of Japanese mechanics is calculated by a very complicated and detailed secret procedure not only based on seniority, as is commonly assumed, but also on the performance of the employee evaluated in categories like discipline, quality of work, quantity of work, responsibility, co-operation with colleagues and superiors, improvement of work, planning and scheduling, and leadership [16]. There is no simple multiplication of the hourly payment negotiated for the currently performed task times the duration of the presence on the shop-floor, as commonly practised in Western factories; there is especially no direct interrelationship between the wages and the actual job contents or job description.

The Japanese intellect has been and still is subject to a generally very emotional and controversial discussion, recently intensified by a flood of Japanese publications about the *Nihonjin-ron*, the "Japanese national character". It is neither appropriate nor sensible to reopen this debate here; however, in the context of illustrating peculiarities in process innovation approaches, a few arguments should still be reviewed briefly.

Most of the current dispute centres around the Japanese capabilites in creativity. One side habitually claims that the Japanese are still (admittedly quite advanced) copying machines, whereas the other admires the "obviously superior" unique and mystical Japanese way of gathering and especially applying new ideas. However that may be, a few characteristics seem to be generally agreed upon: "Japanese thinking" is more phenomenological, more oriented towards concrete signs, manifestations and rules, and less logical and abstract than the German "ratio" in particular (most critically reviewed in the first few sections of this chapter). As a consequence the Japanese intellect is much less occupied by the stubborn pursuit of consistency in some self-determined model of the world, and thus has fewer difficulties in adapting and readjusting itself to the current necessities, whatever these may be (for a more detailed contrasting of the German and the Japanese intellectual structure see [17]).

The interaction of these three phenomena allows the actualization of the unprecedented and extremely efficient automation strategy in the bigger Japanese companies indicated above. The pay policy, by not focusing on the actual job contents, facilitates job rotation or the change of jobs. Along with the "silent co-operation" of the (company-oriented) unions, which care for the prosperity of the firm more than for the well-being of the employees and therefore keep their mouths shut at least in technological questions, reorganizing the shop-floor and reassigning tasks is not anywhere near as difficult in Japan as it would be in the USA or Germany (see also [18]).

Automation in the big companies is further disburdened by the existence of the extensive "buffer" of the small subcontractors. It is neither necessary to automate the manufacturing of an array of parts comprehensively, come hell or high water, nor does it make sense necessarily to automate 100% of anticipated throughput. Parts that are difficult to machine or assemble in FMSs or transfer lines, as well as a small portion of workpieces which are usually machined automatically, may be left for the still relatively cheap and, at the same time, devoted subcontractors to be processed the "old-fashioned" way. This fosters the ties to the subcontractors, further making the most of them in boom periods on the one hand, and, on the other, still allows the mother company to suspend the contracts temporarily and thus keep at least the in-house capacity running in times of economic downturns.

A lot of scholars still attribute the inauguration and especially the functioning of these strategies wholly to smart management – a notion which must be seriously questioned. Although Japanese automation strategies undoubtedly should give everybody some interesting approaches and stunning peculiarities to ponder about, it must be acknowledged that not only the evolution but also the enforcement of these methods is also enabled and supported by a certain set of social and intellectual characteristics.

German workers, for example, may not as easily be appointed to another job in order to improve organizational structures. Despite the fact that through their generally thorough vocational training they are quite comprehensively skilled and therefore imaginative and flexible in their profession, this flexibility withers away when it comes to the application of new technologies. Giving up their at least subconsciously manifested foundation of job-related paradigms is as little their cup of tea as impartially and patiently fiddling about with a tool as stubborn and inscrutable as a computer, be it a numerical control or a computer-aided design (CAD) system. Much more than their Japanese colleagues, German workers have to understand exactly how a new system or structure operates (see also the enlightening comparative study on the introduction of CNC systems by Whittaker [19]), and they have to be convinced that the intended process innovation is workable, necessary, and does not impose any threat to their reputation and expertise. These peculiarities of the "German intellect" are not necessarily bad preconditions for finally mastering huge steps and entirely new approaches in process innovation; however, they certainly complicate the introduction and instant application of new technologies in the short run.

Misunderstood thoroughness is another obstacle to the successful implementation of automation strategies in German companies. Düll and Bechtle [20] in one of their case studies illustrated the inauguration of a newly planned, highly automated assembly factory in a subsidiary of a multinational holding company, where "technosophists" could prevail in their strategy of almost 100% automation even in work planning and scheduling against the rather conservative "economists". Shortly after opening, however, this plant was shut down again. One of the main reasons was the fact that due to the "automated" huge administrative overhead, the assembly cost in the new manufactory was calculated to be still higher than before, even without considering the substantial cost of the implementation.

Obviously, the actualization of process innovation in Germany is not anywhere near as manageable as it is in Japan. This holds true, even though the comprehensive planning of successive steps in the automization process from a variety of perspectives as well as the thorough design of flexible manufacturing or CIM concepts must absolutely be considered a domain of German engineering science. Nowhere else in this world may such a concentration of external institutions and expertise in industrial automation be found, be it the factory-like institutes of the "CIM priests" Spur, Warnecke, or Weck, similarly huge technical university departments, the mutually impairing armada of industrial sociologists, co-ordinating governmental associations, or the technologically active and perceptive metal workers' union IG Metall. But what at first glance looks like an automator's dream may well end up becoming his nightmare: by the time all these self-proclaimed intellectuals may finally have conjoined their divergent perspectives, the Japanese will probably have implemented the fourth generation of a successively improved FMS already.

In the American machine tool industry one hardly dares to touch on the "attitude towards the automation of production". Since machine tool companies became an interesting plaything for financial speculations and attracted the interest of major conglomerates in the late 1960s, the shop-floor has become increasingly alienated and more a subordinate parameter in business calculations than an important integrated part of the production process. In the wake of this development things went from bad to worse; leveraged buy-outs in the late 1970s and early 1980s, potentially a legitimate means of increasing the managers' motivation to show more commitment for the well-being of their firm, were merely utilized to ascertain quick profits in "smart" financial transactions – often by selling and re-renting investment goods, or even by disposing of the whole shop-floor, in order simply to auction the leftovers of the company and again ensure a sound overall profit.

The example of one formerly quite famous American lathe manufacturer should illustrate this situation. Still employing a workforce of 1500 people in 1980, by 1990, a series of mergers and takeovers had caused the number of employees to drop to a remainder of merely 300 scared characters. The shop-floor as well for the most part disappeared, degrading a formerly honourable trendsetter in lathe technology to an insignificant assembly mill utterly dependent on repair jobs or a few loyal customers such as some local factories or the vocational training department of the navy.

A similarly sad story about the extinction in 1985 of the Burg company of Los Angeles is the subject of the formidable book *When the Machine Stopped* [6].

In addition to this own goal scored by exaggerated capitalism, process innovation in American machine tool companies carries still another burden: Taylorism. Even though Taylorism – the separation of planning and execution, of "manual" labour and "intellectual" decision making – is a widespread practice in any industrialized country even nowadays (to be sure, also in Japan), American managers also take this policy to an extreme.

In a study by Salzman surveying the introduction of new technologies on the shop-floor in about 60 of the more important American companies in the metalworking industry [21], only four companies (6%) reported using design policies that were oriented towards utilizing the workers' skills, whereas 22% had policies specifying equipment designs to "lock out" any type of operator intervention. Managers still to a considerable extent seem to be convinced that "Design needs to minimize the 'creativity' of the operator", and that "less human control leads to more production".

Encumbered by a lack of financial resources on the one hand, and the attitude to introduce automation to gain more control over the workforce rather than to increase the efficiency of production (see also [22]) on the other, the degree of automation, at least on the shop-floor of American machine tool manufacturers, is mostly poor. There is, however, a considerable application of CAD systems in the design departments, possibly also due to the strong and influential computer industry in this country. But even here the implementation policy is to lay off the draughtsmen rather than to increase their efficiency. The impression is reinforced that a potentially interesting pragmatic "American way" in process innovation is smothered by a distorted reliance on the self-regulation of capitalism and the centralized organization of Taylorism.

9.4 A Roundabout of Peculiarities – Recursive Reinforcement Stabilizing the Production Styles in Each Country

The discussion of cultural parameters influencing the different stages of the production process must be concluded with one important consideration. Individual peculiarities determining and characterizing the "production style" of a country do not just isolatedly develop within a distinct cultural and social context; to a considerable degree they also reinforce themselves recursively. One example of such a recursion is illustrated in Fig. 9.4, which demonstrates the mutual reinforcement of the "specialization phenomenon" in Germany from education through company orientation to manufacturing and back. The important implication of this example, as well

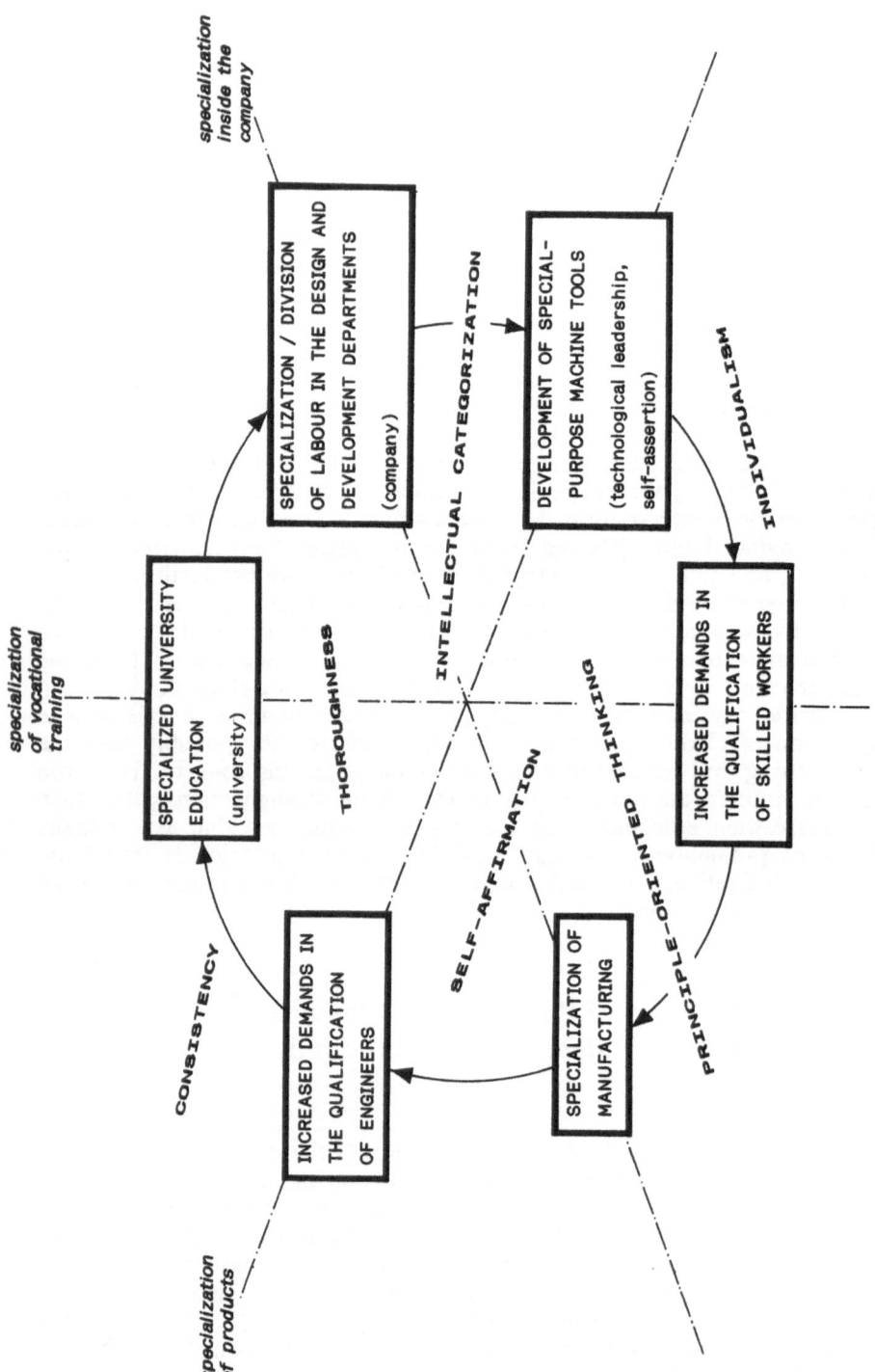

Fig. 9.4. The recursive reinforcement of specialization in German industry.

as of the general denotation of recursion, is the fact that changes in production philosophy, management policies, or even just education guidelines must still be masterminded more carefully. As an integrated course of action these changes may be very successful in gradually redirecting or restabilizing a consistent "culture of manufacturing"; isolated panic "reforms" of individual entities, however, may destabilize the whole "culture" and may just as well lead to a disastrous breakdown of the whole system as to the intended improvements of a few facets.

9.5 Closing Up the Production Process – Often Discussed and Rarely Realized

The comprehensive illustration of only two of the many non-technical parameters influencing, in this case, the design and manufacturing process, should convey one major message: there can neither be an unreflected "picking out" of a few supposedly interesting management practices nor a transfer of a complete management system from one culture to another. The way designers perceive themselves must also play a major part in the discussion about the transferability of design strategies, as the different attitude towards process automation has to be one focus of consideration in any argument about the "shedding" of production.

Yet another aspect in the complex scenario of production styles has still not been touched upon. In a plea for awareness and co-operation, the longitudinal integration of the different stages of the production process must not be neglected (Fig. 9.5). Discussion of the influence of a complex array of factors on individual stages of production should not lead to the conclusion that the respective separation and distinct accentuation of these stages, the chronological or contextual succession, or even the mere existence of some of the stages is a matter of course, a natural phenomenon.

The recent discussion of simultaneous engineering or job rotation has brought this topic widespread recognition; however, as some of the case studies revealed, especially in Germany this aspect has either not been understood or is purposefully misinterpreted. The leader of one of the design departments in a renowned German car company was convinced: "Sure we have simultaneous engineering. We even set up our own 'simultaneous engineering group' which meets once a month and discusses all the related problems". And job rotation? "We have employed job rotation for a long time already. Not between the design and the experimental department, however. Good designers need a lot of years to mature. But within the experimental department, technicians will sometimes work on the hydraulics and sometimes on the electrics".

It goes without saying that at least this section chief does not have the faintest idea what is meant by simultaneous engineering, and what the true purpose of job rotation may be. And this phenomenon is not uncommon in German industry; here the above statements sound much less absurd than they would in Japan. Does this indicate that the longitudinal integration, too, is not only currently one main topic of discussion among engineering scientists, but also subject to influences of non-technical, culture-specific parameters? Probably so. Specialization and dominance of design depart-

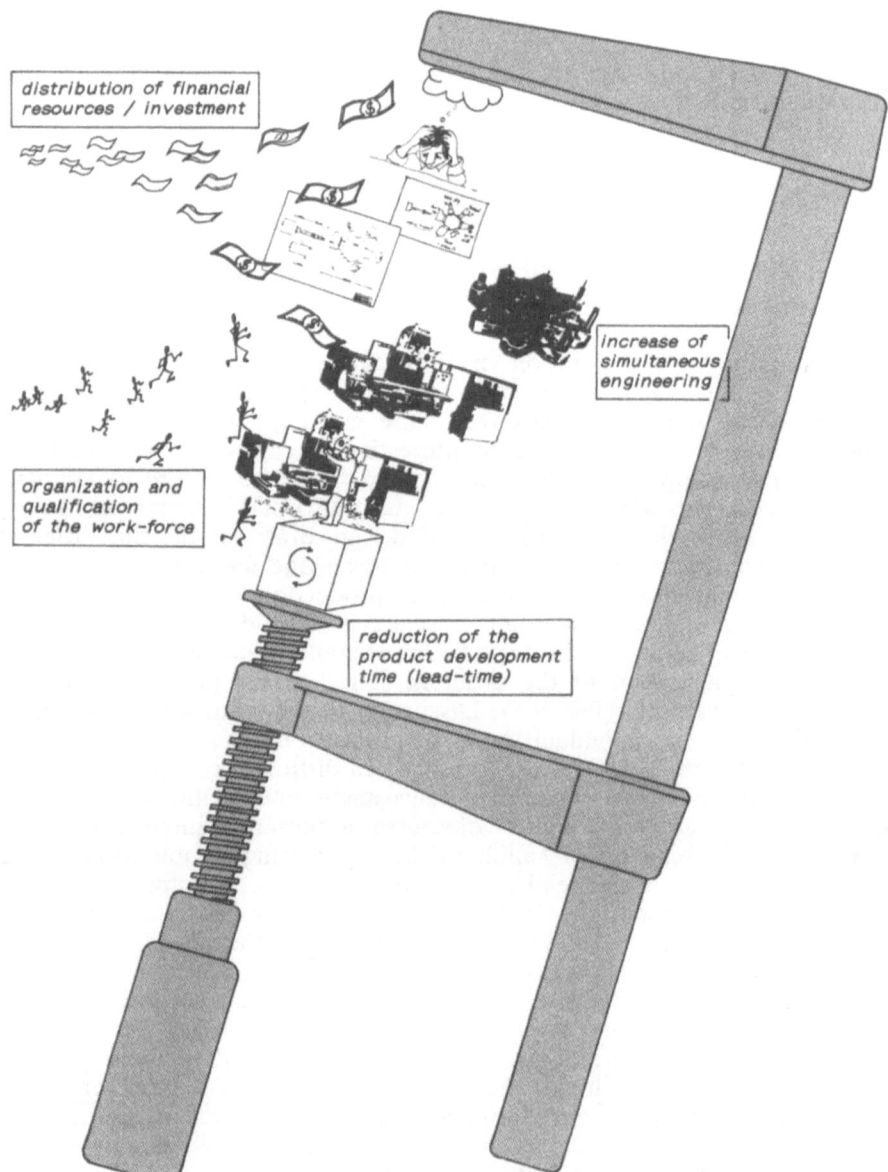

distribution of financial
resources / investment

increase of
simultaneous
engineering

organization and
qualification
of the work-force

reduction of the
product development
time (lead-time)

Fig. 9.5. The longitudinal integration of the production process.

ments as well as the principle- and structure-oriented intellect in Germany not only complicate organizational changes, but also lead to a more vertically segregated production process. The design department will put more emphasis on delivering a thoroughly thought-out, innovative, "perfect" design to the shop-floor, even though "perfect" holds true more in the designer's, not in the production engineer's, frame of reference. Identification with the result of one's own job rather than the final product is important, an improvized preliminary start to manufacturing at least difficult.

In Japan this situation is decidedly different. Identification with the company's goals results not only in fewer reservations regarding organizational changes and thus a more facile planning of processes and manpower, but also in a greater dominance of marketing considerations throughout the workforce. Before the design department has any say in the development of the product, most decisions – even those regarding technical functionalities – are already settled. The designers are "degraded" to a more elaborating function and will have neither the motive, nor the wish, nor the guts to oppose the simultaneous testing of modules by the experimental department or the preparation of manufacturing on the shop-floor.

Financial planning and investment is another subject which has to be regarded in the overall context of production rather than as "naturally" separated into the different stages or departments it has been allocated to. Japanese companies, for example, invest about two-thirds of their R&D funds in process technology, and a mere one-third in product technology; this ratio is almost exactly the reverse in American companies [23]. As a result, manufacturing automation in Japan is quite advanced, and thus the percentage of manufacturing cost in the list price of the product comparatively low. As a recent AAS study demonstrates [15], manufacturing has only a share of 35% to 42% of the overall price in Japanese machine tool companies, but a hefty 53%–57% in the calculation of European producers.[7] Besides the higher degree of automation (also accounted for by the higher proportion of mass production) and the cost-saving potential of the dualistic industrial structure discussed above, however, another reason for this significant difference in cost calculation is the Japanese pricing strategy. Fifteen percent to 35% of the list price is reserved as a discount allowance for dealers and subdealers, considerably more than the meagre 5% allowed for by European producers.

These two examples should terminate the aria on integration and awareness as fundamental approaches to fostering an understanding of what is actually meant by "culture of manufacturing". Obviously this discussion is not anywhere near finished; a lot of aspects listed in Figs 9.2 and 9.5 have not even been touched upon yet. Nevertheless, this may be just the right way to present a case study, as complexity instead of completeness is suggested; a completeness which, as demonstrated in the first few sections, is not comprehensively traceable anyway.

If through the examples and the related discussion the reader has gained an understanding and a "feeling" that production is a complex interplay of technical as well as non-technical concurrently effective and mutually influencing parameters, the main objective of this article has been achieved.

---

[7] This example is cited disregarding the author's strong reservations against equalizing "European" companies in many international comparative studies.

9.6   Doing What Cannot be Done – Contrasting the Product Development Process in the Machine Tool Industry in Japan, the United States, and Germany

According to what has been dramatically beseeched in this whole contribution so far, these concluding sections should not have been written. If production really is an utterly complex entity, it is neither possible nor justified to summarize the product development process in three different countries in just a few paragraphs. However, the following considerations must be seen as a kind of trade-off. Simply insisting on the fact that a production style is an infinitely complex configuration and thus transcends any attempted description, however legitimate this argument may be, leaves a kind of dissatisfied feeling – how then to understand, to organize, to improve?

Therefore in the following three sections the product development patterns in the machine tool industry in Japan, the USA, and Germany will briefly be outlined and some of the respective strengths and weaknesses discussed. However, to qualify this procedure, a few considerations have first to be re-emphasized or clarified.

The following illustrations can at most illustrate a few significant and important characteristics of the respective product development approaches. They can neither be complete, nor can they sufficiently mirror the actual complexity of the process. Thus they must be understood as no more than a simplified attempt to summarize and reflect a few peculiarities of three distinct product development styles confined to the batch production of machine tools.

9.6.1   Synthetic, Modular, Incremental – Product Development in Japan

Figure 9.6 portrays a number of important aspects in the product development pattern of Japanese machine tool companies. It looks all but "lean"; the multiform, hardly decipherable, and diffuse interplay of departments with an unfathomable distribution of responsibilities must rather be regarded as an essential characteristic of the Japanese procedure. Further particularities are the extensive preparatory work done before the actual conceptualization of the machine itself, the comprehensive accumulation of potentially relevant information from all sources, the modular or incremental "short step" approach in product innovation, the important role of testing and prototyping, and the frequently scheduled meetings encompassing experts from the marketing and R&D department all the way down to the assembly people. Few considerations are devoted to tackling eventual technical problems; most prospective difficulties are either "outplanned" or dealt with by thorough and extensive experiments.

In batch or mass production, this procedure is certainly superior to the American or German approach; possibly the design process would deserve some more attention and creative scope to reduce the manpower needed in the earlier product development stages. Very small lot sizes, however, need a considerable adaptation and "slimming" of this complex procedure,

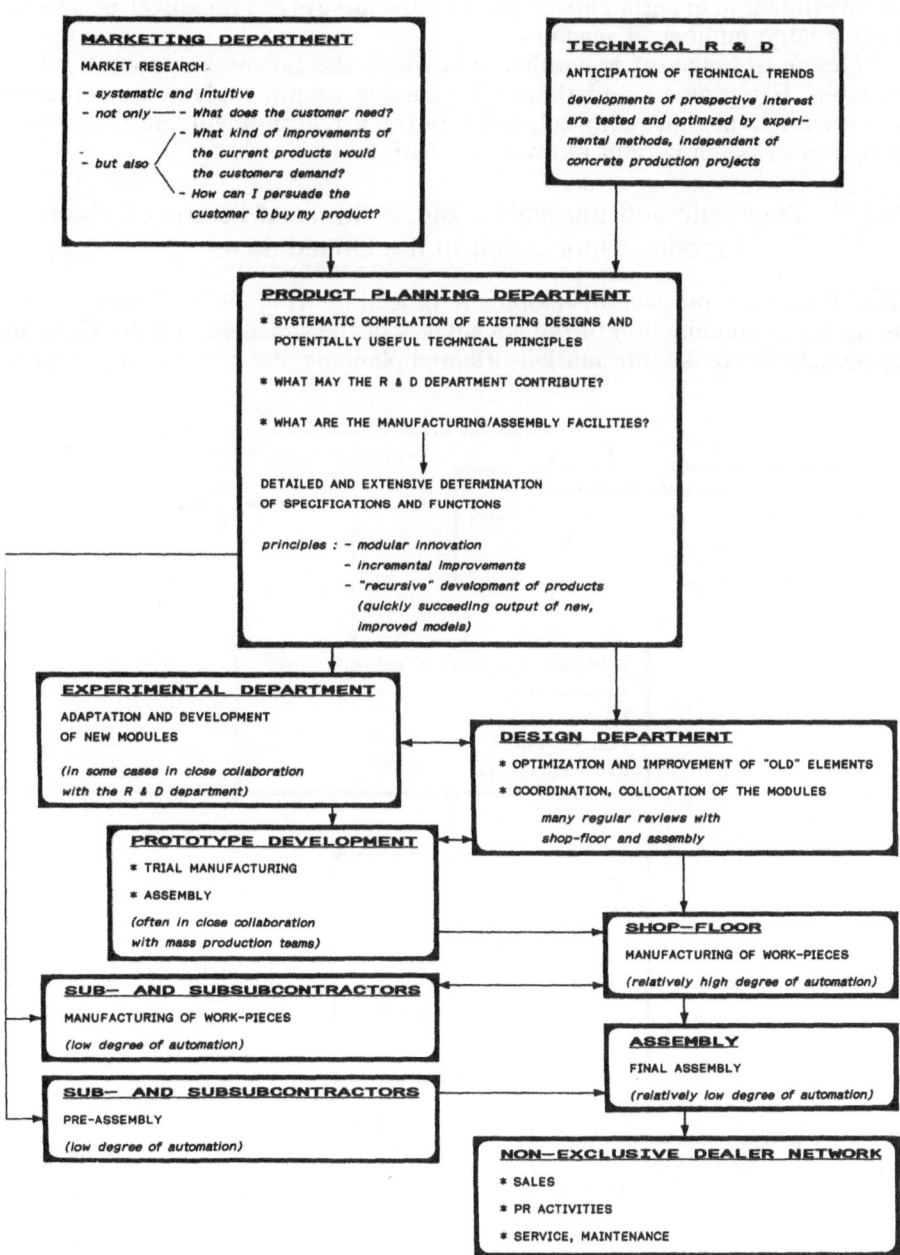

Fig. 9.6. The product development process in the Japanese machine tool industry (batch production).

as investment in experiments or preliminary surveys can no longer be spread over a large number of machines.

"Design to function" is another area where the Japanese approach is less suitable. Experiments and thorough planning cannot replace the formerly indispensable abstract comprehension of the actual problem and the instant proposal of potential and innovative solutions.

9.6.2 Pragmatic and Illimitable, but in the Fatal Pursuit of Short-Term Profits – Innovation in the United States

The American product development procedure (Fig. 9.7) in many ways looks like a combination of the advantages of the Japanese and the German approach. There is more market-oriented planning and a better opportunity

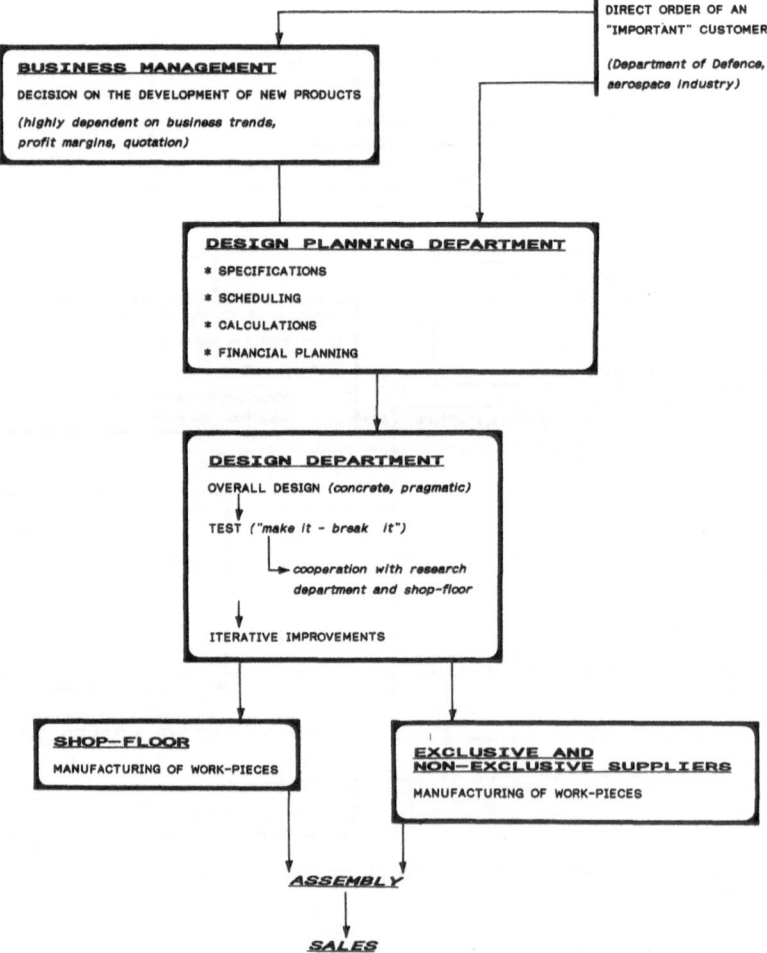

Fig. 9.7. The product development process in the American machine tool industry (batch production).

for pragmatic experimenting in the early stages of the design process, compared with Germany, yet there are more emancipated designers able to acknowledge and digest many viewpoints by unbiased listening and still to decide instantly when needed.

The above section sounds like the author is making fun of the American mechanical engineering industry in light of its recently decayed economic situation. And, sure enough, some people will correctly argue that American employees in no way demonstrate an all-embracing devotion towards their company as do their Japanese fellows, and American engineers and skilled workers are just not as comprehensively educated and intellectually alert as their German counterparts. But these are not the facts that finally make the difference! An unreflected devotion to the company may well have its disadvantages, especially when it comes to abandoning intellectual constraints in technical problem solving and innovative leaps; a misunderstanding of intellectualism may similarly have negative effects by reducing the inventiveness and communication skills to one's own domain.

The American enigma must be looked for somewhere else, unfortunately right in the centre of the American self-understanding. Free capitalism, and free and equal social rights to everybody, are founded on a philosophical misconception, as is the communist ideology of Marx [24, 25]. The pursuit of these beliefs, the conviction that a number of intricate regulations between "fair" and "laissez faire", will solve the inherent contradictions, will for the most part only foster the phenomenon that "some people are more equal than others" (after Orwell).

Industrial strategies based on the social ideology of an accumulation of free and independent individuals who just have to be administered in the right fashion, are obviously not competitive with those based on either the social capitalism of Germany or the feudal capitalism of Japan!

9.6.3   Analytic, Skilled, and Blinded by the Love of Technology – Design in Germany

In the German conception and practice of industrial production (Fig. 9.8), design plays a much more important role, and correspondingly design departments have a significantly stronger position in the company and a more determining influence on product development than in either Japan or the USA. Considering furthermore the network of trendsetting specialized supplier companies and a number of leading scientific-technological institutions it is well justified to talk about the German industry as an "engineering supremacy".

Delighting in the worldwide admiration of their technological expertise, the German engineers unfortunately often neglect a few important realities. A design must not only appear as an ingenious blueprint of conglomerated innovative ideas, but must also function wherever operated; its constituents have to be machined and assembled; and the final product has to be sold, meaning that it has to fit an existing or an arisen need and has to have an affordable price.

The second problem in the functioning of German product development patterns is the misguided perception of specialization. A lot of energy in German companies is wasted as "frictional loss" by insisting on one's own

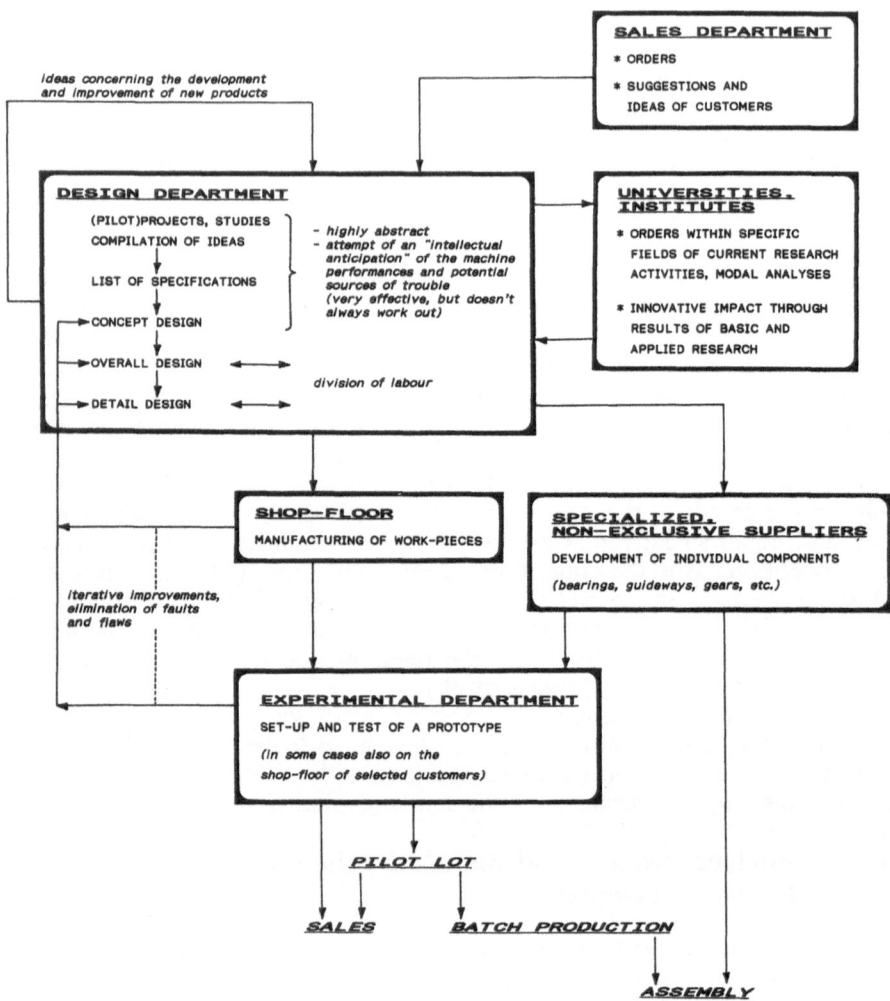

Fig. 9.8. The product development process in the German machine tool industry (batch production).

priorities and expertise, by the intent to increase one's glamour rather than the final product, and by the lack of communicative skills reflected in conversations focused on mutual criticism rather than on understanding each other's problems and viewpoints.

Specialization does not mean segregation. The German engineers and managers seem to be the ones in most desperate need to anticipate and promote this watchword.

References

1. Warnecke H-J. Die Fraktale Fabrik. Springer, Berlin, 1992
2. Altmann N, Sauer D (eds). Systemische Rationalisierung und Zulieferindustrie – Sozialwissenschaftliche Aspekte zwischenbetrieblicher Arbeitsteilung. Campus, Frankfurt New York, 1989
3. Womack JP et al. The machine that changed the world. Rawson Associates 1990
4. Haraguchi E. Toppu shohin o umidasu gijutsusha no niizu tansakuho. ("Needs" searching methods of engineers to produce hit products.) Nikkei Mech 1989; 6(26): 24–36
5. Rosenbrock HH (ed). Designing human-centred technology, Springer, London Berlin Heidelberg, 1989
6. Holland M. When the machine stopped. Boston, Harvard Business School Press, 1989
7. Moritz EF. Die Konstruktionsausbildung in Japan im Vergleich zu Deutschland (Diploma thesis). Lehrstuhl für Konstruktion im Maschinenbau, München, 1988
8. Markus HR Kitayama S. Culture and the self: implications for cognition, emotion, and motivation. Psychol. Rev 1991; 98(2): 224–253
9. Abegglen JC, Stalk G jr. Kaisha. Basic Books, New York, 1985
10. Moritz EF, Ito Y. Computer aided production management – innovation and design. In: Sixth international conference on computer aided production engineering, Edinburgh, 1990
11. Schulz H, Eibeck G. Neue Entwicklungstendenzen im Werkzeugmaschinenbau. Werkst Betr 1964; 97(12): 887–897
12. Maurice M, Mannari H, Takeoka Y, Inoki T. Des entreprises francaises et japonaises face à la mecatronique. Laboratoire d'economie et de sociologie du travail, Aix-en-Provence, 1988
13. Tokunaga S, Altmann N, Nomura M, Hiramoto A. Japanisches Personalmanagement – ein anderer Weg? Campus, Frankfurt New York, 1991
14. Morishima M. Why has Japan succeeded. Cambridge University Press, Cambridge, 1983
15. AAS report to CECIMO. Selling into Japan's machine tool market. Asia Advisory Service KK, Tokyo, 1991
16. Nomura M. Social Conditions for CIM in Japan. International conference on company social constitutions under pressure to change, Berlin, 1990
17. Moritz EF. Konfuzius – Japan – Technik. Ein alter Hut neu aufgesetzt. In: Wissenschaftliches Jahrbuch 1991, Deutsches Museum, München 1992, pp 131–175
18. Jürgens U, Malsch T, Dohse K. Moderne Zeiten in der Automobilfabrik. Springer, Berlin Heidelberg New York, 1989
19. Whittaker DH. Managing innovation: a study of British and Japanese factories: Cambridge University Press, Cambridge, 1990
20. Düll K, Bechtle G. Massenarbeiter und Personalpolitik in Deutschland und Frankreich. Campus; Frankfurt New York, 1991
21. Salzman H. Engineering pespectives and technology design in the United States. AI Soc 1991; 5: 339–356
22. Noble DF. Forces of production – a social history of industrial automation. New York, 1984
23. von Randow G. Den Niedergang aufhalten. Produktion 1991; 28:3
24. Northrop FSC. The meeting of East and West. Ox Bow Press, Woodbridge, 1979
25. Bell D. The cultural contradictions of capitalism. Basic Books, New York, 1976

Subject Index

degree of complexity 155
difficulty of machining 155, 167
Directed graph 153–4
Division of labour, system concepts based on 46–9, 53–6
DNC systems 46, 48
Dynamic scheduling system 88, 103–5
  computer-based 88
  with expert systems 103–5

Education, professional 128–30
  COMETT programme 130
Electronic control stations (ECSs) 35–44
  development trends 42–4
  scheduling and control functions 36–7
  technical and organizational integration 39–40, 62–3
Electronic data processing (EDP) 48
Eye-mark camera 155, 164–6

FA card, material requirement planning 91–3, 96
Factory
  decentralization 112
  FAS tasks 73–4
  future 109
Flexible assembly systems (FAS) 67
  case study 96–107
Flexible automation, development 110
Flexible manufacturing cell (FMC) 1–2, 115
  compactly cubic-like type 5–9
Flexible manufacturing system (FMS) 1–2, 67–9, 114–17
  as office-oriented automation 52–3
  CNC control 115
  definition of 1–2, 68
  degree of automation 54
  design possibilities 52–62
  DNC control 115–16, 125
  for Asian countries 18–19
  hierarchical control structure 116
  information system 58
  installation number and states of 53–4, 67
  machine control 59
  manufacturing system 56
  material flow system 58
  operation of. See FMS operator
  peripherals 60
  shop-floor-(and human-)oriented concepts 59
  skill of worker 10–13
  system components and peripherals 56–60, 68–9
  system design 69
  task model 70–1
  technology 68

three forms of work organization for 54–6
unspecified time 73–4
work design 56–60
work organization 52–6
worker decoupling effect 54, 58–9
worker's task at 73–4
Flexible production islands 62
Flexible transfer line 115
Flexibly computer-integrated manufacturing structure (FCIMS) 1–18, 109–13
  case study (EC countries) 119–31
  case study (Japan) 133–49
  component of 110, 112
  concept of 3
  control and delivery instructions of small parts 140
  core technology 5–10
  data collection from production result and progress control 140
  definition of 112
  developing trend 14–18
  distributed cell type 5
  global network type 2, 13, 18–19, 133–49
  history of 2–4
  human-amenity-secured type 15–17
  human-automated facilities of intermediate type 17
  human intelligence-related function 3
  human-oriented type 15, 90
  interactive correction of production plan 139, 144
  job number management 137
  maturity of 5–8
  MRP procedure implementation 138–9
  MRP production schedule 143–4
  of future 5–9
  parts management 137–8, 146–7
  production master plan 136–9
  production ordering 139–44
  progress management 147–8
  unskilled worker-oriented type. See Unskilled worker-oriented manufacturing
  VTR cylinder workshop analysis 136–7
FMC. See Flexible manufacturing cell (FMC)
FMS. See Flexible manufacturing system (FMS)
FMS operator (operation) 69–71, 74–80
  creative resource-level 72, 74
  models of attitude of 71
  perceptual-cognitive resource-level 72, 74
  sensory-motoric resource-level 72, 74
Future production environment 5–9, 30
  possible products 19–20
  possible system configurations 5–7
  roles of human being 9–12

Gantt chart 35, 139
Group work 32